New Insights in Herbicide Science

Edited by Kassio Ferreira Mendes

Published in London, United Kingdom

New Insights in Herbicide Science
http://dx.doi.org/10.5772/intechopen.100871
Edited by Kassio Ferreira Mendes

Contributors
Kassio Ferreira Mendes, Daniel Resende Fontes, Andrew De Paula Ribeiro, Marcelo Rodrigues Dos Reis, Miriam Inoue, Mohammad Taghi Alebrahim, Elham Samadi Kalkhoran, Te-Ming Paul Tseng, Ermias Misgansw Amare, Mahdi Banaee, Islam R. Ghoniem, Suman Bagale

First published in London, United Kingdom, 2023 by IntechOpen
IntechOpen is the global imprint of INTECHOPEN LIMITED, registered in England and Wales, registration number: 11086078, 5 Princes Gate Court, London, SW7 2QJ, United Kingdom

British Library Cataloguing-in-Publication Data
A catalogue record for this book is available from the British Library

Additional hard and PDF copies can be obtained from orders@intechopen.com

New Insights in Herbicide Science
Edited by Kassio Ferreira Mendes
p. cm.
Print ISBN 978-1-80356-107-3
Online ISBN 978-1-80356-108-0
eBook (PDF) ISBN 978-1-80356-109-7

Meet the editor

Kassio Ferreira Mendes is a Professor of Biology and Integrated Management of Weeds, at the Department of Agronomy, Federal University of Viçosa, Brazil. He graduated from the State University of Mato Grosso in 2011 and received a master's degree in Agronomy (Crop Science) from the Federal University of Viçosa in 2013. He obtained a doctorate and post-doctorate degrees in Sciences - Nuclear Energy in Agriculture (Chemistry in Agriculture and Environment) from the Center of Nuclear Energy in Agriculture, Luiz de Queiroz Campus, University of São Paulo, Brazil, in 2017 and 2019, respectively. He completed research fellowships at the Department of Soil, Water, and Climate; College of Food and Agricultural Sciences; University of Minnesota, USA; and the United States Department of Agriculture - Agricultural Research Service (USDA-ARS). Dr. Mendes is a member of the Brazilian Society of Weed Science.

Contents

Preface

The first synthetic herbicide, 2,4-dichlorophenoxyacetic acid (2,4-D), was discovered in 1942 and its effectiveness and selectivity caused a paradigm shift in weed management practices in agriculture worldwide. Since then, the chemical control of weeds has become the most widely used method in the world because it is efficient, relatively low cost compared to other control methods, easy to use, and professionally appropriate.

However, most growers have an immediate and economic view of weed control, which can create environmental problems in the medium and long term. Repeated applications of herbicides with the same mechanism of action have been common practice in many parts of the globe. Frequently used chemical weed control coupled with target site mutation has led to the emergence of many herbicide-resistant weed biotypes reported worldwide.

This herbicide resistance of weeds is undoubtedly a major concern in modern agriculture, and some mechanisms of action present greater resistance problems than others. The largest number of cases of resistant biotypes belong to the mechanism of action of acetolactate synthase (ALS) inhibitors, followed by photosystem II (PSII) inhibitors. Today, herbicides with new mechanisms of action are necessary to control the evolution of resistance of these biotypes to existing herbicides.

In addition, it is essential that herbicides are properly applied alone or in the mixture to preserve the final quality of the harvested products as well as the natural resources that sustain production, especially soil and water. The correct way to apply these products aims for maximum biological efficacy and minimum damage to neighboring crops, the environment, and humans. The more suitable the equipment and techniques employed, the greater the efficacy. In this sense, herbicide application technology involves economically depositing the correct amount of the biologically active product on the target at the right time in the required quantity with minimal environmental contamination.

In view of this, it is important to study the behavior of herbicides in plants and soil for weed control. This involves estimating the trends to which herbicides are subject, as a function of three main processes: retention, transformation, and transport in the soil. Thus, the mere fact that an herbicide reaches the plant's leaves or is applied to the soil where it develops is not enough for it to exert its action; it must penetrate the plant, translocate, and reach the organelle where it will act. To determine the correct product to be used in each crop and in each soil, the professional must know the product's metabolism characteristics and the plant's sensitivity to the product and/or its metabolites, among other factors, such as the most promising form of application.

New Insights in Herbicide Science discusses these issues in six chapters. It is a useful resource for undergraduate and graduate students, technicians, professors, farmers, and all those involved in this area.

Kassio Ferreira Mendes
Professor of Biology and Integrated Management of Weeds,
Department of Agronomy,
Federal University of Viçosa,
Viçosa, MG, Brazil

Section 1

Application History, Mode of Action and Resistance

Chapter 1

Liquid Chromatography Tandem Mass Spectrometry after the QuEChERS Method for Determining 20 Herbicide Residues in Wheat and Flour

Islam R. Ghoniem

Abstract

Agriculture is the backbone of the economy and social structure, and it plays a critical part in each country's overall growth. Because of the significant food gap that exists in several vital crops, wars, and the continual expansion in the population, the role of agriculture products has recently become critical. The world is currently experiencing a severe food shortage, estimated to be over 60% of its strategic food requirements. As a result, there is a need to increase the area of farmed land in order to satisfy the growing population and raise food demand by eliminating weeds that can reduce agricultural output. Weed is an unwanted plant (one that grows in the incorrect area) that reduces crop output. Herbicides are a type of pesticides that are used to kill weeds and increase crop output. As a result, herbicide residues on food, particularly cereals, must be determined. In this study, the QuEChERS approach for determining herbicides in wheat and corn by direct injection to Exion HPLC coupled with a SciexQtrap API 6500+ LC–MS/MS system using an electrospray positive ionization (ESI+) at lower concentrations without utilizing acids or clean-up is evaluated, optimized, and validated in this work.

Keywords: QuEChERS, LC–MS/MS, agriculture, herbicides, cereals

1. Introduction

As a result of the continuing expansion in the world's population, the use of pesticides in contemporary agriculture has become one of the most critical necessities for meeting society's food needs, and millions of tons of pesticides are used annually for this purpose [1]. Pesticides are one of the most commonly utilized substances on the planet. Despite their usefulness, pesticides are one of the most dangerous compounds that damage humans, animals, and surface water in particular [2]. When pesticides are used in large quantities in the environment, they have the potential to harm the environment, especially human health [3]. Weeds are any unwanted plants that grow in a field and threaten crops, animals, or human health. Herbicides are a type of pesticide that kills weeds to protect plants and boost crop output [4]. Herbicides are frequently employed in agriculture and turf management in the landscape. They account for almost 70% of all agricultural pesticide use worldwide [5]. Herbicides can cause everything from skin rashes, nausea, and weariness to headaches, chest pain, and even death in some cases.

Pesticides are used in roughly 2 million tons over the world, with 47.5% being herbicides, 29.5% being insecticides, 17.5% being fungicides, and 5.5% being other pesticides [6]. China, the United States, Argentina, Thailand, Brazil, Italy, France, Canada, Japan, and India are the top ten pesticide-using countries in the world [7]. Furthermore, it is predicted that by 2020, global pesticide usage will have increased to 3.5 million tons [8]. Africa's economy is heavily reliant on agriculture, with approximately 59% of the population relying on it for a living [9]. Despite this, the African continent contributes 2–4% of the global pesticide market share and has the lowest pesticide usage rate in the world [9]. Food demand is expected to rise rapidly in the next three decades as a result of the rising population, and demand for pesticides, herbicides, and fungicides are also expected to rise [10].

The quick, easy, cheap, effective, rugged, and safe (QuEChERS) approach was used to detect this chemical and estimate its concentration [11–18]. In terms of analysis costs and turnaround time, multiresidue methods are the most efficient way for herbicide analysis. The majorities of the procedures have multiple steps and use a lot of different solvents and reagents. In terms of good recovery, short duration of analysis, cheap cost, and safety, the QuEChERS approach combined with liquid chromatography–tandem mass spectrometry (LC–MS/MS) was determined to be the optimal combination for determining herbicides in some foods. Because of the more ionized herbicides, LC–MS/MS is now commonly employed [12–14, 19, 20].

Controlling herbicide residues in food items through monitoring and a maximum residue limit (MRL) setting is critical for consumer safety. The Codex Alimentarius Commission (CAC) and the European Commission determined MRLs based on residues in food that must be found at safe levels for consumers [21, 22]. In the European Union (EU) legislation, the lowest limit of analytical quantitation (LOQ) is specified as the MRL that equals 0.01 mg/kg if the MRL obtained by different trials is not safe for consumers [22].

Yingying et al. [23] improved and validated a QuEChERS technique for determining florasulam and pyroxsulam residues in wheat grain and straw using liquid chromatography–tandem mass spectrometry (LC–MS/MS). The approach was tested on cereals such as oat, millet, corn, and rice. Average recoveries ranged from 76 to 113%, with RSDs ranging from 2 to 15%. TAO et al. [24] developed an efficient method for determining various phenoxy acid herbicide residues in grains. The study of phenoxy acid herbicides in rice, corn, and wheat was optimized using a QuEChERS approach combined with high-performance liquid chromatography–tandem mass spectrometry (HPLC–MS/MS). Renata Raina et al. [25] developed pesticide residue testing procedures for a wide range of foods, including cereal-based foods, nutraceuticals and associated plant products, and infant feeds. Many processed consumer products are made from these grain, fruit, vegetable, and plant-based components. A modified QuEChERS approach has been applied for cereal and nutraceuticals, which are dry sample products, with additional steps to allow wetting of the dry sample matrix and subsequent cleanup using dispersive or cartridge format SPE to eliminate matrix effects.

Wheat is a widely cultivated crop whose seed is a grain that is consumed as a staple food all over the world. The most important wheat types are common wheat (*Triticum aestivum*), durum wheat (*Triticum durum*), and club wheat (*T. aestivum*) (*T. compactum*). Wheat is grown as a commercial crop because it generates a high yield per unit area, thrives in a temperate climate with a short growing season, and produces versatile, high-quality flour. Wheat flour is used to produce bread, pasta, cereal, pastries, cookies, crackers, muffins, tortillas, and pitas, among other things. Wheat is the second most widely grown cereal grain after maize, and its global trade volume exceeds that of all other crops combined. The total global wheat production in 2020 was 760 million tons. China, India, and Russia are the world's three greatest individual wheat producers, accounting for over 41% of global wheat production. Individually, the United States is the world's fourth-largest wheat producer. If the European Union were counted as a single entity, it would produce more wheat than any other country save China [26].

The current study's technique describes the examination of a mixture of herbicides in various matrices after extraction using the QuEChERS technology. The QuEChERS technique is evaluated, optimized, and validated for the determination of 20 herbicides in wheat and flour by direct injection to LC–MS/MS at lower concentrations without the use of acids or clean-up in this study. Exion HPLC paired with the SciexQtrap API 6500+ LC–MS/MS System was used to determine these chemicals utilizing electrospray positive ionization (ESI+).

2. Experimental method

2.1 Instrumentation and analysis

1. LC–MS/MS system, ExionLC AC coupled with Qtrap API 6500+ MS/MS system from AB Sciex, USA.

2. Chromatographic column, Infinity lab Poroshell 120 EC-C18 3.0 × 50 mm, 2.7 μm particle size (Agilent, USA).

The injection volume was 2 μL and the column temperature was 40°C. The pesticides are separated using a Gradient mixing program of 10% 50 mM ammonium format in deionized water, which is mostly used for positive ionization mode, with 0.1% formic acid as eluent A and methanol as eluent B at 300 μL/min flow rate starting by A bottle 60% for 1 min, changed continuously till 11.5 min to be 10% for 0.5 min, changed progressively till 12 min to be 0% for 2 min and returned to 60% from A in min 14 for 2 min to be 16 min complete run time for every one of the 20 pesticides. Electrospray ionization in the positive ion mode with multiple reactions monitoring (MRM) mode was used to complete the MS/MS analysis.

The LC mobile phase stock solution was 50 mM ammonium formate solution in methanol/water (1:9), and the LC mobile phase was 10 mM ammonium formate solution in methanol/water (1:9), dilute 200 mL of LC mobile phase stock solution with 800 mL methanol/water (1:9), adjust the pH to about 3.78 ± 0.02 with ammonia solution (33%), and then add 100 mL methanol and LC mobile phase was 10 mM ammonium formate solution in methanol/water (1:9), dilute 200 mL of LC mobile phase stock solution with 800 mL methanol/water (1:9), the pH should be 4 ± 0.1, adjust as needed.

2.2 Reagents and materials

Atrazine (99%), clodinafop (free acid) (99%), clodinafop-propargyl ester (99%), cycloxydim (98.8%), diphenamid (99%), fenoxaprop-P-ethyl (R-enantiomer) (99%), haloxyfop-2-ethoxyethyl ester (99%), haloxyfop (free acid) (99%), imazamethabenz-methyl (97.4%), imazethapyr (99%), mesosulfuron-methyl (98%), metolachlor (98.5%), metribuzin (99.5%), metsulfuron-methyl (99.5%), pendimethalin (98.8%), quizalofop-ethyl (99.3%), quizalofop-P-ethyl (98.4%), simazine (98%), sulcotrione (99%), and triclopyr butotyl (99.1%) were purchased from Dr. Ehrenstorfer (Augsburg, Germany). Methanol (99.9%) HPLC grade was purchased from J.T. Baker (PA, USA). Acetonitrile 99.9% -HPLC grade was purchased from J.T. Baker (Pennsylvania, USA). Deionized water (<18M_cm resistivity) was performed in the laboratory using a Millipore (Billerica, MA, USA) MilliQ water purification system. Ammonia solution (33%) was purchased from Riedel-de Häen (Seelze, Germany). Formic acid (98–100%) was purchased from Riedel-de Häen. QuEChERS extraction kits.—5982–5650 was purchased from Agilent {Agilent QuEChERs salts and buffers are prepackaged in anhydrous packages(4 g MgSO4; 1 g NaCl; 1 g trisodium citrate dihydrate;0.5 g disodium citrate sesquihydrate)} (Santa Clara, CA, USA).

2.2.1 Standard preparation

Stock solutions (1000 μg/mL) of each pesticide standard were prepared by dissolving atrazine in toluene, clodinafop (free acid) in toluene, clodinafop-propargyl ester in toluene, cycloxydim in toluene, diphenamid in toluene, fenoxaprop-P-ethyl (R-enantiomer) in toluene, haloxyfop-2-ethoxyethyl ester in toluene, haloxyfop (free acid) in toluene, imazamethabenz-methyl in toluene, imazethapyr in methanol/toluene (3:7 v/v), mesosulfuron-methyl in toluene/acetone (7:3 v/v), metolachlor in toluene, metribuzin in toluene, metsulfuron-methyl in toluene, pendimethalin in toluene, quizalofop-ethyl in toluene/acetone (8:2 v/v), quizalofop-P-ethyl in toluene, simazine in acetone, sulcotrione in toluene/acetone (9:1 v/v), and triclopyr butotyl in toluene/acetone (9:1 v/v). All stock solutions were prepared and kept at −20 ± 2°C. Working mixtures of the examined pesticides (5 g/mL each) and calibration mixtures of concentration levels 0.01, 0.05, 0.1, and 0.5 g/l were made by diluting suitable aliquots of the stock solutions with methanol kept at 4 ± 2°C.

2.2.2 Spiked samples preparation

The flour and wheat were purchased at the local market. The samples were thoroughly ground before being homogenized in an electric mill. In recovery experiments, wheat and flour samples were spiked with a suitable amount of working mixture standard solution.

2.3 Extraction procedure

Herbicide residues in wheat and flour were extracted using the QuEChERS technique for herbicide residue analysis. Initial single-phase extraction of 2 g of homogenized sample with deionized water in a 50 mL PFTE centrifuge tube, 10 mL deionized water added, tube closed and shaken vigorously by geno grinder at 500 rpm for 1 min, and then with acetonitrile in a 50 mL PFTE centrifuge tube, 10 mL acetonitrile added, tube closed and shaken vigorously by geno grinder at 500 rpm for 1 min.

DOI: http://dx.doi.org/10.5772/intechopen.104964

After that, a mixture of Agilent QuEChERs salts and buffers is added to the tube, which is then closed and rapidly shaken for 1 min at 500 rpm with a geno grinder, then centrifuged for 5 min at 4000 rpm (3430 rcf). The cleaned extract is filtered using syringe filters (0.45 m) and transferred to a PP vial after centrifugation. Finally, the liquid sample was injected into a liquid chromatography-mass spectrometry (LC–MS/MS) apparatus.

3. Result and discussion

The analysis technique used in this study was created with the goal of detecting and quantifying as many herbicides as feasible in a single run. When deciding which herbicides to include, two criteria were used: (1) herbicides registered for crop protection by local authorities, and (2) searching the literature for commonly studied compounds. Acidification was used in this method in the form of buffer citrate salt (trisodium citrate dihydrate and disodium citrate sesquihydrate), which served two purposes: (1) improving extraction by converting conjugate of some herbicide to neutral form, thereby increasing recovery, and (2) adjusting pH 5–5.5, thereby increasing herbicide sensitivity. The herbicides were determined using LC–MS/MS with an ESI source and MRM mode, which offered a highly selective and sensitive technique. All of the target analytes were ionized to $(M + H)^+$ form in the positive mode, according to the physicochemical parameters of the target. The positive mode was chosen since it works well for the majority of analytes. Herbicides can be quantified directly using the LC–MS/MS approach, which does not require any derivatization and requires minimal cleaning. A QuEChERs approach was used to design the method for 20 herbicides. The chromatograms obtained for each compound, as shown in **Figure 1**, were determined with sufficient precision and accuracy. The approach was tested on a total of 20 herbicides, each with a distinct retention time of 16 min. Although an excellent summary of the LC–MS/MS methods used for herbicides was offered, it did not cover all herbicides discussed in this study, and only a few studies for determining several classes of herbicides in wheat and flour in a single multiresidue approach were published.

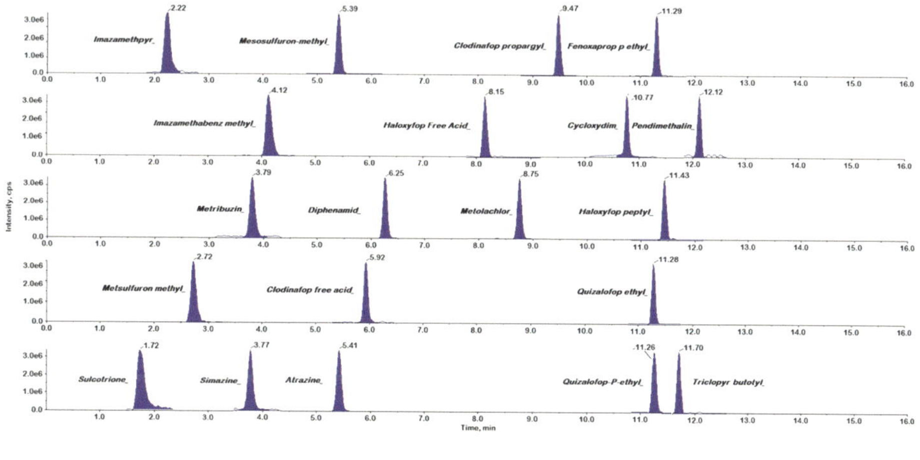

Figure 1.
The approach was validated using chromatograms produced by liquid chromatography tandem mass spectrometry (LC–MS/ MS) with electrospray ionization (ESI) in positive mode and MRM mode for 20 herbicides used in the study.

3.1 Mass spectrometry study of 20 herbicides

To discover the best precursor, product ions, and operating conditions, 20 herbicides were injected directly into the LC–MS/MS system in 1:1 methanol at a concentration of 0.1 μg/mL. **Table 1** summarizes the precursor and product quantification and confirmation ion pairs, as well as the declustering potential and collision energies.

3.2 Method validation

The developed method was validated in compliance with the document's method validation standards SANTE/2020/12830 document [27].

3.2.1 Linearity of calibration curves

Plotting the detector response area ratio vs. the concentration of the analytical solutions at various concentration levels ranging from 0.001 to 0.1 μg/mL established the linearity of the calibration curve of 20 herbicides. The calibration curves were prepared using sex levels of calibration standards in the concentration ranges of 0.001, 0.002, 0.005, 0.01, 0.05, and 0.10 μg/mL. Plotting the peak area vs. concentration yielded a calibration curve. According to European guidelines, the analytes showed linear behavior in the studied concentration levels with a correlation

No.	Acidic herbicides	Q1	DP	Q3	EP	CE	CXP
1.	Atrazine	216.1	82	104	10	37	10
2.	Clodinafop (free acid)	312	41	237.9	10	33	4
3.	Clodinafop-propargyl ester	350	115	266	10	24	10
4.	Cycloxydim	326.3	61	280	10	19	16
5.	Diphenamid	240.1	31	134.1	10	25	4
6.	Fenoxaprop-P-ethyl (R- enantiomer)	362.1	71	288.1	10	23	7
7.	Haloxyfop-2-ethoxyethyl ester	434	92	288	10	49	10
8.	Haloxyfop (free acid)	362	81	316	10	25	18
9.	Imazamethabenz-methyl	289	117	161	10	37	10
10.	Imazamethpyr	290	81	177	10	41	18
11.	Mesosulfuron-Methyl	504	76	182	10	33	10
12.	Metolachlor	284.2	76	252.2	10	21	4
13.	Metribuzin	215.1	81	187.2	10	21	4
14.	Metsulfuron-methyl	382	76	167	10	21	6
15.	Pendimethalin	282	88	194	10	25	14
16.	Quizalofop-ethyl	373	120	299	10	20	10
17.	Quizalofop-P-ethyl	373.1	71	298.9	10	25	15
18.	Simazine	202.2	77	131.9	10	27	8
19.	Sulcotrione	329	86	111	10	39	10
20.	Triclopyr butotyl	356.2	122	237.7	10	15	14

Q1: Precursor ion, Q3: Product ion, DP = Decluster Potential [V], EP = Entrance Potential [V], CE = Collision Energy [V] and CXP = Collision Cell Exit Potential [V].

Table 1.
List of herbicides and MRM parameters in LC-MSMS-ESI positive mode.

Herbicide	R^2
Atrazine	0.9961
Clodinafop (free acid)	0.9982
Clodinafop-propargyl ester	0.9963
Cycloxydim	0.9997
Fenoxaprop-P-ethyl (R- enantiomer)	0.9986
Haloxyfop-2-ethoxyethyl ester	0.9992
Haloxyfop (free acid)	0.9991
Imazamethabenz-methyl	0.9976
Imazamethpyr	0.9966
Mesosulfuron-Methyl	0.9977
Metolachlor	0.9969
Metribuzin	0.9991
Metsulfuron-methyl	0.9999
Pendimethalin	0.9998
Quizalofop-ethyl	0.9987
Quizalofop-P-ethyl	0.9965
Simazine	0.9995
Sulcotrione	0.9978
Triclopyr butotyl	0.9999

Table 2.
R^2 values for the 20 herbicides.

coefficient (r2) greater than 0.99 as shown in **Table 2**, indicating that all analytes were within the acceptable range and the coefficient of variation (CV percent) for each calibration point was less than 20% [28].

3.2.2 Matrix effect

A matrix effect research was carried out on blank wheat and flour samples using a conventional herbicide mixture of 20 herbicides. To correct for matrix-induced suppression in LC–MS/MS, matrix-matched standard calculations were performed at 0.01, 0.05, and 0.1 mg/kg.

The following formula was used to make the calculations:

$$\text{Matrix effect \%} = ((\text{peak area STD in matrix/peak area STD in solvent}) - 1/100).$$

To compensate for the matrix effect suppression on the results, 450 μL of blank sample was fortified with 50 μL of 0.5 μ g/mL standard solutions to achieve 0.05 μg/mL concentration levels [29].

3.2.3 Quantification limit (LOQ)

The quantitation limit of all of the substances investigated was determined to be 0.01 mg/kg for all of them. The validity of this level has been established in accordance with the SANTE guidelines [28] and EU 396/2005 regulation [22].

Compound	n	Wheat (0.01 mg/kg)		Flour (0.01 mg/kg)		Wheat (0.05 mg/kg)		Flour (0.05 mg/kg)		Wheat (0.1 mg/kg)		Flour (0.1 mg/kg)		Reproducibility
		Rec.%	CV%	Rec.%	CV%	Rec.%	CV%	Rec.%	CV%	Rec.%	CV%	Rec.%	CV%	Pooled CV%
Atrazine	6	85	5	79	4	89	3	97	4	101	7	95	6	4
Clodinafop (free acid)	6	80	7	77	6	91	2	93	3	94	7	96	7	5
Clodinafop-propargyl ester	6	80	10	82	5	95	2	99	8	103	7	82	7	5
Cycloxydim	6	78	13	78	6	98	2	98	12	106	6	95	5	11
Diphenamid	6	82	9	80	4	89	2	96	6	99	7	89	7	8
Fenoxaprop-P-ethyl (R-enantiomer)	6	79	9	79	7	93	2	97	8	103	7	84	7	5
Haloxyfop-2-ethoxyethyl ester	6	83	9	81	5	96	2	96	3	101	7	93	7	5
Haloxyfop (free acid)	6	84	8	80	5	91	2	93	3	99	7	96	6	7
Imazamethabenz- methyl	6	86	12	79	5	95	2	98	2	99	7	96	7	3
Imazamethpyr	6	83	11	85	15	83	2	94	3	102	7	83	8	4
Mesosulfuron-Methyl	6	87	9	82	3	89	2	103	6	102	6	98	6	4
Metolachlor	6	83	13	79	10	90	2	96	8	105	7	80	7	7
Metribuzin	6	82	7	71	5	93	2	98	3	101	7	116	6	4
Metsulfuron-methyl	6	83	8	82	4	96	2	103	10	98	5	88	6	4
Pendimethalin	6	82	15	74	14	86	2	91	9	104	6	103	5	8
Quizalofop-ethyl	6	80	15	82	5	95	2	100	10	103	7	81	7	6
Quizalofop-P-ethyl	6	80	13	81	5	93	2	100	13	104	7	74	7	5
Simazine	6	84	13	79	5	93	2	97	4	101	7	92	7	4
Sulcotrione	6	102	17	96	16	94	2	79	13	100	6	95	10	9
Triclopyr butotyl	6	90	6	82	7	90	2	93	6	97	7	85	7	8

n: No. of replicates, Rec: Mean recovery, CV: Coefficient of variation.

Table 3.
Average recoveries and coefficient of variation (CV%), on wheat and flour samples were spiked at 3 different concentration levels 0.01, 0.05 and 0.1 mg/kg.

3.2.4 Accuracy and precision

Six replicate spiked wheat and flour samples were analyzed at three distinct levels (0.01, 0.05, and 0.1 mg/kg) to acquire accuracy and precision. The percentage of the money recovered ranged from 71–105%. The precision was based on the corresponding relative standard deviations, and the trueness was based on the mean recoveries (RSD). **Table 3** shows the recoveries, means, and RSD percent. Reproducibility (interday accuracy and precision) was tested over a two-month period at a fortification level of 0.05 mg/kg and found to be less than 12%.

4. Conclusion

The current study developed a multiresidue technique of testing for 20 herbicides with a limit of determination of 0.01 mg/kg, which meets the EU MRLs for wheat and flour farm goods. Two MRMs for quantification and conformation were chosen based on the optimal declustering potential and collision energy, and the mass spectrometric parameters were tuned to give the best sensitivity. In terms of approved recovery, short duration of analysis, cheap cost, and safety, the QuEChERS method followed by Exion HPLC and a SciexQtrap API 6500+ LC–MS/MS system using an electrospray positive ionization (ESI+) technology was shown to be the optimal combination for determining the 20 herbicides. Herbicides can be quantified directly using the LC–MS/ MS method, which does not require any derivatization and requires minimum cleanup with a total runtime of 16 min. The majority of the chemicals tested had recovery rates ranging from 71–105%, with relative standard deviations of less than 12%, indicating adequate precision. Recovery trials on six replicates of spiked blank wheat and flour samples at 0.01, 0.05, and 0.1 mg/kg were used to determine the method's precision and accuracy. The developed assay was linear over a concentration range of 0.01–0.5 μg/mL, with a correlation coefficient of more than 0.99 at the 0.01 μg/mL limit of quantification.

Acknowledgements

The author thanks the Central Laboratory of Residue Analysis of Pesticides and Heavy Metals in Foods, which was awarded an international Accreditation Certificate in all analyses by the Finnish Branch (FINAS) of the European Accreditation Center for Laboratories (EAL).

Authors' contributions

The study's inception and design were aided by the author. Islam R. Ghoniem was in charge of material preparation, data collecting, and analysis. Islam R. Ghoniem wrote the first draught of the manuscript and provided feedback on prior draughts. The final manuscript was read and approved by the author.

Funding

Not applicable. The author works at the Central Laboratory of Residue Analysis of Pesticides and Heavy Metals in Foods, which supported this research.

Data availability

On reasonable request, the corresponding author will provide the datasets used and/or analyzed during the current work.

Declarations

The author declares no competing interests.

Author details

Islam R. Ghoniem
Central Laboratory of Residue Analysis of Pesticides and Heavy Metals in Foods, Agricultural Research Center, Giza, Egypt

*Address all correspondence to: islam_refaat@qcap-egypt.com

References

[1] Rani M. Studies on decay profiles of quinalphos and thiram pesticides [Ph.D thesis]. Uttarakhand, India: Indian Institute of Technology Roorkee; Chapter 1, 5; 2012

[2] Mcknight US, Rasmussen JJ, Kronvang B, Binning PJ, Bjerg PL. Sources, occurrence and predicted aquatic impact of legacy and contemporary pesticides in streams. Environmental Pollutant. 2015;**200**:64-76

[3] Poulier G, Lissalde S, Charriau A, Buzier R, Delmas F, Gery K, et al. Can POCIS be used in Water Framework Directive (2000/60/EC) monitoring networks ? A study focusing on pesticides in a French agricultural watershed. Science of the Total Environment. 2021;**497-498**:282-292

[4] Acidic Herbicides Definition. Available from: http://ec.europa.eu/food/plant/pesticides/index_en.htm [Accessed: 2022]

[5] van den Brink PJ, Mann RM. Impacts of agricultural pesticides on terrestrial ecosystems. Ecological Impacts of Toxic Chemicals. 2011:63-87

[6] Carvalho FP. Pesticides, environment, and food safety. Food Energy Secur. 2017;**6**:48-60. DOI: 10.1002/fes3.108

[7] Worldatlas. 2018. Available from: https://www.worldatlas.com/articles/top-pesticide-consuming-countries-of-the-world.html

[8] Zhang W. Global pesticide use: Profile, trend, cost / benefit and more. Proceedings of the International Academy of Ecology and Environmental Science. 2018;**8**:1-27

[9] Abate T, Van Huis A, Ampofo JKO. Pest management strategies in traditional agriculture: An African perspective. Annual Review of Entomology. 2000;**45**:631-659. DOI: 10.1146/annurev.ento.45.1.631

[10] Snyder J, Cairns Smart J, Goeb J, Tschirley D. AGRICULTURAL RESEARCH INSTITUTE OF MOZAMBIQUE Directorate of Training, Documentation, and Technology Transfer Research Report Series Republic of Mozambique Pesticide use in Sub-Saharan Africa: Estimates, Projections, and Implications in the Context of Food; 2015

[11] Anastassiades M, Lehotay SJ, Štajnbaher D, Schenck FJ. Journal of AOAC International. 2003;**86**:412

[12] Chung SWC, Chan BTP. Journal of Chromatography A. 2010;**1217**:4815. DOI: 10.1016/j.chroma.2010.05.043

[13] Koesukwiwat U, Sanguankaew K, Leepipatpiboon N. Analytica Chimica Acta. 2008;**626**:10. DOI: 10.1016/j.aca.2008.07.034

[14] Pareja L, Cesio V, Heinzen H, Fernández-Alba AR. Talanta. 2011;**83**:1613. DOI: 10.1016/j.talanta.2010.11.052

[15] Rebelo AM, Dolzan MD, Heller M, Deschamps FC, Abate G, Micke GA, et al. Journal of Brazilian Chemical Society. 2016;**27**:186

[16] Santilio A, Girolimetti S, Barbini DA. Food Additives Contamination Part A. 2014;**31**:845. DOI: 10.1080/19440049.2014.891296

[17] Santilio A, Stefanelli P, Girolimetti S, Dommarco R. Journal of Environmental Science and Health Part B. 2011;**46**:535

[18] Yang F, Bian Z, Chen X, Liu SS, Liu Y, Tang G. Journal of AOAC International. 2013;**96**:1134. DOI: 10.5740/jaoacint.12-467

[19] Raina-Fulton R. Journal of AOAC International. 2014;**97**:965. DOI: 10.5740/jaoacint.SGERaina-Fulton

[20] Wang J, Chow W, Cheung W. Journal of Agricultural Food Chemistry. 2011;**59**:8589. DOI: 10.1021/jf202158g

[21] Codex Alimentarius Commission pesticides MRLs. 2019. Available from: http://www.fao.org/faowho-codexalimentarius/standards/pestres/pesticide-detail/en/?p_id=31 [Accessed: 2022]

[22] EU Pesticides Database DG Sanco EU MRLs Regulation EC No. 396/2005. 2018. Available from: http://ec.europa.eu/sanco_pesticides/public/index.cfm [Accessed: 2022]

[23] Bi Y, Han L, Song S, Yao W, Qin F, Xu Y, et al. Method validation, storage stability and field trial for residues of florasulam and pyroxsulam in cereal by liquid chromatography with tandem mass spectrometry. Food Additives and Contaminants: Part A. 2020;**37**:793-803. DOI: 10.1080/19440049.2020.1723809

[24] Guo T, Wang X, Wang H, Hu Y, Zhang S, Zhao R. Determination of phenoxy acid herbicides in cereals using high-performance liquid chromatography-tandem mass spectrometry. Journal of Food Protection. 2019;**82**:1160-1165. DOI: 10.4315/0362-028X.JFP-18-558

[25] Raina-Fulton R. New trends in pesticide residue analysis in cereals, nutraceuticals, baby foods, and related processed consumer products. Journal of AOAC International. 2015;**98**:1163-1170. DOI: 10.5740/jaoacint.SGE2Raina-Fulton

[26] Wheat Production by Country 2022. Available from: https://worldpopulationreview.com/country-rankings/wheat-production-by-country [Accessed: 2022]

[27] SANTE Guidance Document on Pesticide Analytical Methods for Risk Assessment and Post-approval Control and Monitoring Purposes. Sante/2020/12830 2021, 1-50

[28] SANTE/11945/2015, Method Validation and Quality Control Procedures for Pesticide Residues Analysis in Food and Feed. European Commission, Directorate-general for health and food safety, Brussels, 2015

[29] Kwon H, Lehotay SJ, Geis-Asteggiante L. Variability of matrix effects in liquid and gas chromatography-mass spectrometry analysis of pesticide residues after QuEChERS sample preparation of different food crops. Journal of Chromatography A. 2012;**1270**:235-245. DOI: 10.1016/j.chroma.2012.10.059

Chapter 2

Modes of Herbicide Action

Suman Bagale

Abstract

Weed Management is one of the most important crop intervention practice to counter crop loss. Different physical, mechanical, biological, and chemical methods are employed for the successful management of weeds. Among these chemical weed management practices focus on managing weeds using several chemical formulations which are commonly known as herbicides. Herbicides control the weed species through interference, mitigation, and disruption of the biochemical and physical processes of a cell. When herbicides are applied to a plant, it gets absorbed through plant surfaces and gets translocated to the specific site of action where it produces toxicity in the physiological and biochemical processes and ultimately check the growth and development of plant species. The sequential process from the introduction of herbicides to till it kills a plant is known as herbicides modes of action. The herbicides modes of action can be studied on nine different headings where the chemical group present in each herbicide acts on specific sites and interferes with the normal functioning of such sites ultimately checking the growth and development of a plant. This chapter is aimed at decoding the specific herbicide action in relation to its chemical family, translocation, action mechanism, and injury produced in the weed species.

Keywords: chemicals, glyphosate, herbicides, herbicide resistance, novel modes

1. Introduction

The global demand for food crops is rapidly increasing with the increase in the world's population, on other hand production of crops is constrained by several factors such as weeds pest, insects, and diseases, among these all weeds are one of the major factors that can cause loss of productivity of field crops. Weeds are any plants that are grown in undesirable places and compete with crops plants for nutrients, sunlight, moisture, and other growth factors. Anything that grows in unintended places are generally referred to as weeds. According to Gharde [1], weeds are notorious yield reducer than pests, disease, and insects, which are thought to cause an estimated loss of 11billion USD in 10 major crops, which causes 31.4% loss in soybean, 30.8% in green grams, 25.3% in maize, 21.4% in mustard, 18.6% in wheat and 21.4% in direct-seeded rice. Reduction of crop yield in crops is due to competition between crops and weeds for space and other growth factors. Yield loss of crops due to weeds depends on several factors such as weed emergence time, weed density, types of weeds, competition ability of crops, and if left uncontrolled, they can cause 100% loss in crop production [2]. The successful and strategic management of weeds can decrease the yield loss significantly, which can ensure more grain harvest. The management of weeds has become one of the most researched aspects in the field of crop science. In small farm size, it can be managed through hand weeding or mechanical weeding machines like cono-weeder and weed-roller whereas its management in the large farms has become a problematic issue. Mostly in the case of commercial cultivation weeds species are mostly managed by using different pre-emergence and post-emergence herbicides. With the increase in the weed resistance towards these herbicides, there is a need for weed science research focusing on herbicide resistance and herbicide mode of action. In a study carried out by Heap [2], it was observed a total of 511 unique cases of herbicide resistance belonging to 266 weed species (153 dicot and 133 monocots) have been reported globally out of which major herbicide-resistance weed species were reported in wheat followed by maize, rice, soybean, spring barley, and cotton. Herbicide-resistant weed populations are rapidly evolving as the process of natural selection and development of traits by weeds to escape the action of herbicides. The graphs show that, herbicide resistance has been steeply increasing from 7 cases in 1975 to 509 cases at the end of 2020. The major herbicide resistant traits were observed in the weed family belonging to Poaceae or grass. The five major weed families Poaceae, Asteraceae, Brassicaceae, Amaranthaceae, and Chenopodiaceae account for 70% of total herbicide resistance cases though they only include 50% of total principal weeds [3]. More weed species are resistant to ALS inhibitors, with the reported 160 species, which is followed by Photosystem II inhibitors. Glyphosate one of the most common post-emergence herbicide used as broad broad-spectrum control of weeds has become less effective due to intensive use of herbicide leading to the quick emergence of glyphosate-resistant biotypes [4]. Mitigating herbicide resistance has become one of the most important things to consider during crop production. The herbicide resistance in plants can be somehow coped with by introducing different herbicides of the varied mode of action, crop rotation, and using integrated weed management practices in crops. The successful management of herbicide resistance in input-intensive agriculture can be combated by diversifying the herbicide products, cultivating crops with combined herbicide resistance, increasing reliance on pre-emergence herbicides than post-emergence herbicide, breeding weed-competitive crop cultivars, and advances in site-specific and precision weed management [5].

The advancement in the field of genetics, plant physiology, chemistry, and plant science has made open to many researchers to understand the basis and mechanism of herbicide resistance. Herbicide resistance mechanisms can be target site resistance, non-target site resistance, cross-resistance, and multiple resistance [6]. The target site herbicide resistance is due to the mutation in genes encoding herbicide enzymes, non-target herbicide resistance is due to the reduced amount of herbicide active ingredients through reduced absorption or translocation. The cross-resistance is due to the use of several herbicides with the same mode of action and multiple resistance is due to two or more herbicide resistance mechanisms in response to a sequential selection of herbicides with a different mode of action. The herbicide mode of action explains how the active ingredients present in commercial herbicide formulation act on plants. The mode of action of herbicide is variable based on the chemical composition of their active ingredients and the weeds species in which they act on. Some herbicides act on plants through the root system, some act on photosynthesis and photosystem, and some herbicides are found to act on the cell membrane and enzymatic pathways. Understanding the mode of action of herbicides is important for the management, classification, organization, and hierarchy of the herbicides as it also provides an insight into herbicide resistance, which has become a problem in sustainable agricultural management [7].

Herbicide enters into a plant system through several different mechanisms. These acting mechanism differs in between the herbicide in relation to the chemical nature that is present in the active ingredients of the herbicide. The herbicide mode of action discusses on the sequence of events from the introduction of herbicide in the environment till its kills the plant through toxicity produced by the chemicals presents in the active ingredients of the herbicide, whereas the herbicide mechanism of action discusses on physiological and biochemical changes caused by the herbicide within plant system. Understanding the mode of herbicide action helps to relate the chemistry of herbicide and the physiology that exists within a plant. The knowledge of acting mechanisms helps to cope with the problem of herbicide resistance and helps to maximize the efficacy of herbicide during weed management. The study incorporates the parts of how herbicide gets absorbed in the plant surface and how they act on the physiology of weed plant and injure them to eliminate them from the competition with crop species. The knowledge on general chemistry, plant physiology, genetics, and plant science can help to decipher the roles that lie beyond the herbicide mode of action. In general herbicides are classified as pre-emergence and post-emergence herbicides. In pre-emergent herbicide the mode of action is principally through absorption from root zones, whereas in the case of post-emergent herbicide the mode of action is mainly through absorption from foliar parts. In general, to acts as an herbicide on a plant, it must pass through certain sequential stages of contact, absorption, movement, toxicity, and death of weed species and the mode of herbicide action to produce injury includes inhibition, disruption, interruption, and mitigation of regular growth of weed species [8]. The exploration of the mode of herbicide action is dynamic and new modes of action of herbicide has been constantly adding, which is helping for the discovery of new herbicide. Based on the site of action of herbicide and mode of action altogether 22 types of herbicide action have been developed. Therefore, understanding the mode of herbicide action can substantially help in understanding the mechanism of herbicide resistance and exploring new strategies to cope with herbicide resistance. So, the chapter focuses on different herbicide mode of action in relation to their chemical family, mechanism of action, translocation, and toxicity. In recent years, a perennial weed of *Roegneria* genus commonly known as wild rye, which is widely distributed in China has shown tolerance to ACCase inhibitor herbicides like fenoxaprop, clodinafop and pinoxaden [9]. It was observed that the ACCase activity were increased by 1.46 and 1.34-fold in wild rye and wheat plant after 72 hours of fenoxaprop treatment than at 0 hrs of treatment as shown in **Figure 1**. It was suggested that the enhanced activity of ACCase is due to enhanced metabolism of herbicide, leading to herbicide tolerance.

2. Lipids synthesis inhibitors

These kinds of herbicides are those which cause disruption in lipid synthesis and check the growth of plants through rupture of the cell wall and cell oozing. The herbicide having group 1 site of action falls under these categories, where herbicide inhibits Acetyl CoA Carboxylase (ACCase) enzyme which is required for fatty acids synthesis that forms a part of phospholipid bilayer in the cell membrane of plant cells. The inhibition of (ACCase) enzymes restricts the formation of cell wall in meristematic regions and ultimately kills the plant cell. The (ACCase) inhibitors herbicides are used for the selective control of weed species, which are found to have resistance with glyphosate herbicide [10]. The mechanism begins when the herbicide comes in

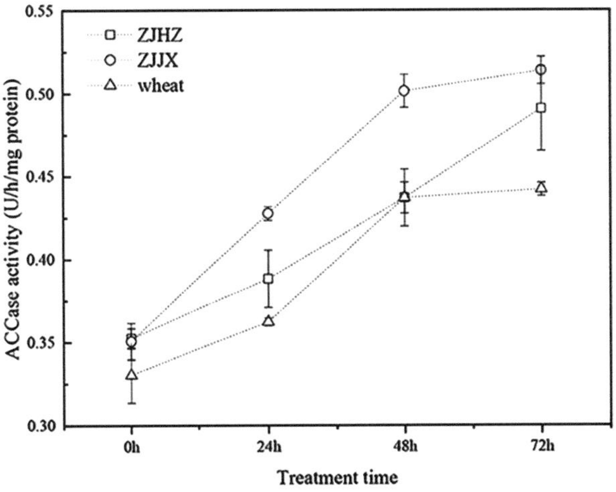

Figure 1.
Increase in the enzymatic activity of ACCase with increase in treatment time in wild oat and wheat plant depicting tolerance to fenoxaprop [9].

contact with plant species and it gets translocated in the meristematic region through phloem where it inhibits the meristematic activity producing necrotic symptoms in the growing tissues after one week of application [10, 11]. The chemical family of this herbicide includes aryloxyphenoxylpropinate, cyclohexanedione, and phenylpyrazole which are applied as post-emergence herbicides to control grassy weeds in broadleaf crops [11]. The common herbicides include fenoxaprop, fluazifop, diclofop, quizalofop, clethodim, sethoxydim, and Pinoxaden. These groups of herbicides are applied through foliar spay and translocated through phloem in meristematic regions. The major injury includes plants turn brown, chlorotic symptoms can been seen in the leaves, and vein browning and purpling can be seen after 3-4 days of herbicide application. (ACCase) inhibitor herbicides are short-lived in soil, relatively low solubility in soil, and used relatively in low rates. They have low leaching potential and are found to be less hazardous to the environment. (ACCase) inhibitors herbicides are found to have resistance against 43 grass weeds species [12].

3. Amino-acid synthesis inhibitors

These kinds of herbicide are found to have two sites of action. They belong to group 2 Acetolactate Synthetases (ALS) that catalases the synthesis of branched-chain amino acids, such as leucine, isoleucine, and valine. The inhibition of these enzymes restricts the biosynthesis of these amino acids, which are the essential part of protein necessary for cell membrane formation. The ALS inhibitors are found to have effect on the reproduction of some plant species such as inducing male sterility and their potency, which can act extremely at low concentrations, and the rapid evolution of resistance to these herbicides in some plants [13, 14]. They are the largest group of herbicides that are post-emergence selective in nature. The chemical ingredients are of these herbicides are absorbed through roots and foliage and translocated through both xylem and phloem. The major injury of ALS inhibiting herbicide includes interveinal chlorosis, purpling and root pruning. The major chemical family includes sulphonyl urea, Imidazolinone, Sulfonylurea, and Triazolopyrimidine [15]. The common

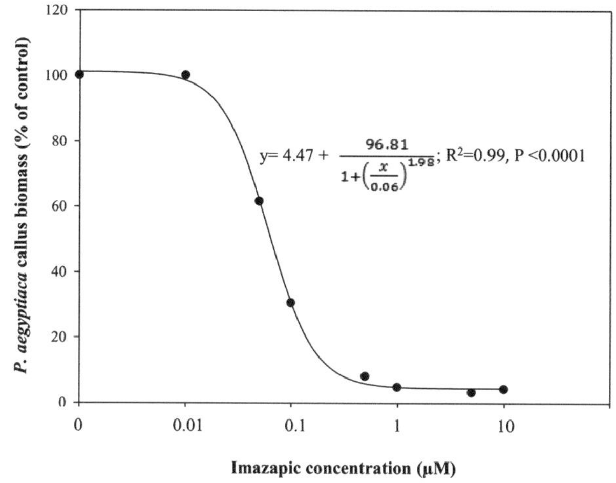

Figure 2.
The control of callus biomass is more effect in lower concentration than in higher concentration of amino acid inhibitor herbicide imazapic [16].

herbicides include imazamox, imazapic, imazaquin, imazethapyr, nicosulfuron, metsulfuron, triasulfuron, chlorsulfuron, rimsulfuron, prosulfuron, pyroxsulam, diclosulam, and flumetsulam. In a study conducted by Dor [16], the tissue culture of broomrape was found to be more sensitive to imazapic in which a concentration of 0.05μM significantly decreased the biomass and a concentration of 10μM caused blackening of died callus, which suggests that free amino acid content increased with the increased in the concentration of imazapic as shown in **Figure 2**.

The second group of herbicides that causes amino-acid synthesis inhibition are group 9 herbicides, which causes blockade in the production of enzymes from 5- enoylpuruvyl Shikimate-3-phosphate (EPSP) pathway. The enzymes in EPSP pathway catalyze the biosynthesis of aromatic amino acids like phenylalanine, tyrosine, and tryptophan. These amino are essential for protein synthesis and the absence of them causes cell membrane disintegration. The broad-spectrum herbicide glyphosate belongs to the chemical family which checks the EPSP pathway [17]. Herbicides having this mode of action is non-selective and absorbed through phloem tissues. They produce major injury in foliage, causing foliage discolorations stunting and killing the plant ultimately. The growth and development of the plant is check right after the herbicide application; the major symptoms appear only after a few days of application.

4. Growth regulators

Growth regular mode of mechanisms of herbicide checks the growth of plant by modulating the balances of growth hormones and regulators within the plant system. The herbicide having this mode of action belongs to two different group of the site of action. Which consists of group 4 herbicides which are generally synthetic auxins such as 2,4-D. The synthetic auxins imbalance the Indole Acetic Acid level and causes growth abnormalities in plants and leading plants to ultimate death. The major chemical family includes Phenoxy, benzoic acids, and carboxylic acids. The common herbicides in use are 2,4-D, 2,4-DB, Dichlorprop, MCPA, MCPB, Dicamba,

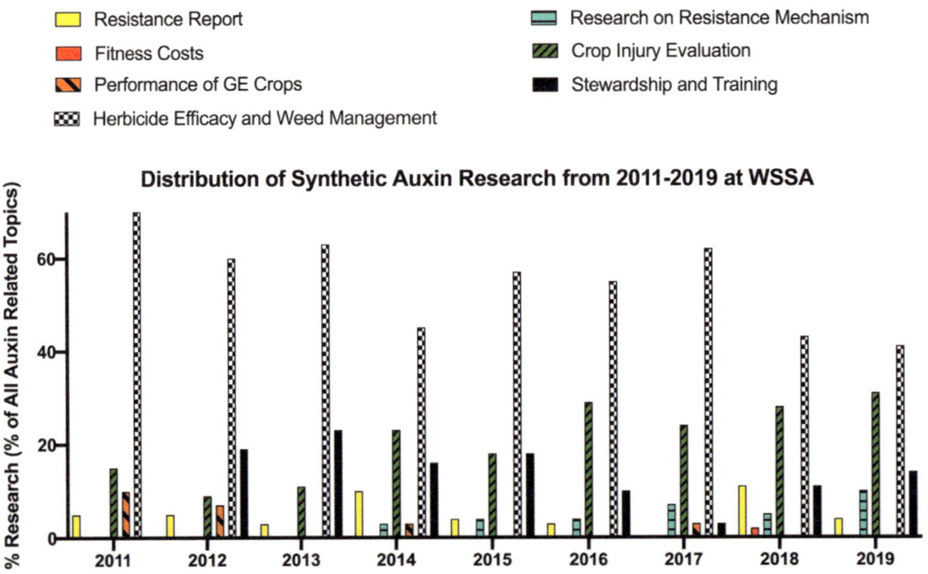

Figure 3.
Distribution of Synthetic Auxin Research from 2011 to 2013. Adapted from Todd [19].

Clopyralid, and Picloram. These herbicides are commonly used to control broadleaf weeds in plant species, having narrow leaves as post-emergence herbicides [18]. The action of herbicide is controlled by multiple factors rather than a single factor, which disturb the nucleic acid metabolism and cell wall integrity. During recent years, the herbicide efficacy and use of synthetic auxin herbicides has been decreased due to the problem of herbicide resistance and the evolution of other herbicides. **Figure 3** depicts the research status on synthetic auxin herbicides from 2011 to 2019 published by WSSA [19].

The group of herbicides belonging to the growth regulators mode of action are chemicals that check the transport of auxin in the meristematic regions. Through the mode of action of these herbicides remain elusive, the majority of these classes of herbicide are found to check the bi-directional flow of auxin by inhibiting vesicle trafficking in plants [20]. The major chemical family of this herbicide are semicarbazone which checks the growth of broadleaf weed in grass crops. The herbicide is absorbed through roots and foliage and they translocate through xylem and phloem. The application of these herbicide during pre-post emergence gives better control of broad leaf weeds. They produce injuries in growth and reproduction abnormalities, leaf malformation, cupping of leaves, abnormal outgrowths of tissues, brittleness in stem, and stalk.

5. Photosynthesis inhibitors

Photosynthesis inhibitors disturbs the process of photosynthesis by binding with the specific binding sites in photosystem II present in the chloroplast of plant cells. Inhibition of photosynthesis could result in slow starvation of the plant and cessation of starch translocation; however rapid death occurs perhaps from the production of secondary toxic substances. Herbicides of photosystem II belong to the following chemical classes: s-triazines, triazinone, uracil, urea, phenyl carbamates, anilide, cyanophenols, dinitrophenol, which are classified into three different groups 5, 6, and 7 on the basis of site of action [21]. The commonly used herbicides having this mode of action are atrazine, simazine, metribuzin, hexazinone, terbacil, bromoxynil, bromacil, pyrazone, bentazon, diuron, and linuron. The group 5 herbicides inhibit photosynthesis by binding within serine in PSII and are absorbed through roots and shoots and translocated through xylem and phloem, group 6 herbicide inhibits photosynthesis by binding with histidine, these herbicide acts as post-emergence contact herbicide, so through spraying of herbicide is recommended. The group 7 herbicides bind with protein complex present in the thylakoid membrane, which checks the transport of electron in the Electron transport Chain. The blocking of electron causes reduced carbon dioxide fixation and production of ATP and NADPH2, which are known as energy packets of respirations. These herbicide controls both narrow and broadleaf weeds. The action of these herbicides is greater during the daytime when there is full sunlight as the herbicide gets activated in presence of light. The herbicides show symptoms of chlorosis and necrosis of leaf margins which progresses towards the base of the leaves after a few hours of application.

6. Nitrogen metabolism inhibitors

This mode of mechanism belongs to herbicide of group 10 having site of action at glutamate synthesis pathway. These herbicide inhibits the production of glutamate syntheses enzymes, which is essential for the conversion of ammonia to other nitrogenous compounds [22]. The blocking causes the accumulation of ammonia ions in the plant leading to increase in PH of the surrounding tissues. This causes protein disintegration, breakage of fatty acids, rupturing of cells, and overall imbalance of ion within cell sap. The major chemical family of herbicides having this mode of action are Phosphorylated Amino Acids commonly traded in the chemical name of glufosinate. These are the broad-spectrum, postemergence herbicide having limited translocation within plant systems so that through spraying of this herbicide is recommended for maximum efficiency. The major injury produced by this herbicide is foliar injury in the plant. The injury symptoms are more prevalent in the younger leaves, in contrast to the deficiency symptoms and plant stress symptoms.

7. Pigment inhibitors

Pigment inhibitors are those herbicides that cause blocking in pigment formation such as anthocyanins, carotene, retinol, and chlorophyll. These herbicides belong to group 12 site of action which blocks the enzymatic activity of 4-hydrooxy phenyl Pyruvate dehydrogenase (HPPD), which plays a role in the synthesis of pigments like chlorophyll, anthocyanin, and carotene. The another group comprises of herbicides from group 13, which causes inhibition of determine synthesis that causes inhibition of synthesis of retinol and degradation of phytin pigments [23]. The major chemical family of these herbicides includes Pyrazole, Pyrazolone, and Pyridazinone. The commonly used herbicides are amitrole, clomazone, isoxaflutole, and mesotrione. The level of pigments is highly reduced leading presence of unbound lipid radicles which causes lipid oxidation, make some protein dysfunctional, and ruptures of the cell membrane. The injury produced by these herbicides is prominent as they show white or bleaching coloration right after the application. In prolonged symptoms expression of translucent leaves, and rapid wilting of weeds species can be seen in the applied area.

8. Cell membrane disrupter

These herbicides interfere with the cell membrane activity, causing them to distort in their structure and functions. The site of action of these herbicides comprises of herbicides belonging to group 14, which causes inhibition of Protoporphyrinogen Oxidase (PPO) enzymes which catalyzes the conversion of ProtoporphyrinogenIX (PPGIX) to Protoprophyrin (PPX). The accumulation of PPGIX causes interbonding to form triplet PPGIX which in the presence of light can disrupt the hydrogen bond, break the bond between fatty acids, and degradation of protein structures. Likewise, triplet PPGIX can obstruct the biosynthesis of chlorophyll and haeme pigments [24]. The chemical family under this group of herbicides are Diphenyl Ether, Thiadiazole, Triazolinone, and Trifluro Methyl Uracil, which includes commonly used herbicides like lactofen, oxyfluorfen, acifluorfen, fomesafen, flumiclorac, and sulfentrazone.

The other group of herbicides, which acts as cell membrane disrupters are chemical belonging to Group 22, which causes inhibition of Photosystem I during photosynthesis. The major chemical family of this group are bipyridylium, which comprises of commercialized herbicides such as diquat and paraquat. These chemical causes the diversion of the electron from the PSI and generate herbicide radicals, which on reacting with oxygen form hydrogen peroxide and hydroxyls radical that causes the breaking of unsaturated fatty acids, chlorophyll, lipids, and proteins in the cell membrane [25]. These herbicides are post-emergence herbicides that get activated under bright light and have a contact mode of action. These herbicides are found to control weeds well under the maturity period too. The major injury system appears in the plant after 1–2 hours of application with evident water-soaked foliage, browning, and necrosis.

9. Seedling root growth inhibitors

These group of herbicides belong to group 3 of the site of action which inhibits the root development in young seedlings by interfering with the cell wall microtubules. Due to this mode of action of herbicides they are commonly called microtubules inhibitors. These chemicals inhibit cell division and cause the blocking of root growth and extension due to the assembly of herbicide-tubulin complex inside microtubules. The complex inhibits the polymerization of microtubules disturbing root cell wall formation [26]. These herbicides are used as pre-emergence herbicides their application through direct soil incorporation gives the best result. The chemical family of these herbicides is dinitroaniline, which is commercialized in the chemical form of pendimethalin. Other commonly used herbicides include trifluralin, ethafluralin, cycloate, and butylate. The major injury of this herbicide is swollen coleoptile, swollen hypocotyl, callus formation, brittle stem, and formation of short secondary roots.

10. Seedling shoot growth inhibitors

Seedling shoot growth inhibitors are the herbicides belonging to group 8 site of action, which interfere with the activity of lipid synthesis through chemical thiocarbamate. These chemical inhibits the biosynthesis of protein, fatty acids, flavonoids, and gibberellins. The other group of herbicides acting as seedling shoot growth inhibitors is long chain fatty acids inhibitors. These herbicide conjugates with Acetyl CoA to form thiocaramate sulfoxide which inhibits long-chain fatty acids during seedling seed growth [27]. The chemical family chloroacetamide, which comprises of chemical herbicides like alachlor, butachlor, and metolachlor are herbicides belonging to this mode of action. They are used as pre-emergence herbicides, soil incorporation of these herbicides give better efficiency. These herbicides are volatile in nature, which are absorbed through roots and emerging shoots and only translocated through xylem vessels in plants. The major injury produced by these groups of chemicals are stunting and enlarged cotyledons.

11. Some novel modes of herbicide action

The problem of herbicide resistance has led the researcher to explore on new modes of herbicide action. Exploring herbicides with a new mode of action can potentially be effective for those weed species which are resistant to conventional herbicides. Several methods are being employed to explore herbicide that acts on new site of action. Major focus has been put on the exploration of phytotoxic primary and secondary metabolites such as protoporphyrin IX and sphingoid bases. The next approach commonly used for the study is identifying the potential site of action with very low-level enzyme level [28]. The herbicide with the target site, Dihydrodipiconitae Sythetase (DHDPS), which catalase the first and rate-limiting step in lysine synthesis is found to be effective in *Arabidopsis thaliana* conformed by using high throughout the chemical screen. The class of inhibitors are found to bind with the novel and unexplored packets within DHDPS, which produces the symptoms of retarded growth and germination [29]. Another novel herbicide with the chemical form tetflupyrolimet belonging to Weed Science Society of America (WSSA) group 28 is found to be effective against the control of long-season grass weeds in rice fields. The herbicide acts on homogentiosata solanesyltransferase (HST) and dihydroorotate dehydrogenase (DHODH) inhibition [30]. The use of tetflupyrolimet has been under research on several other crops like sugarcane, wheat, soybean, and corn. The herbicide is expected to launch commercial in the year 2023, especially recommended for transplanted and direct-seeded rice.

12. Conclusion

Herbicides have become one of the indispensable parts of commercial agriculture to control the weed species efficiently. The continuous and excessive use of herbicides has evolved the problem of herbicide resistance in many weed species. The new exploration of herbicide mode of action has provided new insights on the target action of herbicides and their acting mechanism along with providing solutions for herbicide resistance. A sound knowledge on the mode of herbicide action help farmer to select the herbicide based on degree of weed infestation, a suitable method for herbicide application, and understand the action mechanism involved to check the growth of weeds in field crops. The study of the mode of action interrelates with the study of the site of action, active chemical involved, and injury produced by such chemicals in the growth and development of weed crops which is equally useful for herbicide formulation. Hence, from the discussions in this chapter, it is evident that different herbicides have their own mode of action to kill the field weeds. Knowing the herbicide mode of action can help for the tactical management of weed species and cope with the resistance trait that lies within plant species.

Acknowledgements

The author acknowledges Mr. Roshan Subedi, Assistant Professor of the Institute of Agriculture and Animal Science, Tribhuvan University for his constant guidelines and support during the preparation of this book chapter.

Conflict of interest

The authors declare no conflict of interest.

Author details

Suman Bagale
Institute of Agriculture and Animal Science, Tribhuvan University, Kirtipur, Nepal

*Address all correspondence to: sumanbagale74@gmail.com

References

[1] Gharde Y, Singh PK, Dubey RP, Gupta PK. Assessment of yield and economic losses in agriculture due to weeds in India. Crop Protection. 2018;**107**:12-18

[2] Chauhan BS. Grand challenges in weed management. Frontiers in Agronomy. 2020;**1**:3

[3] Heap I. The international survey of herbicide resistant weeds. Weedscience. 2019;**12**:5

[4] Ferreira Mendes K, Nogueira de Sousa R, Flávia Souza Laube A. Current approaches to pesticide use and glyphosate-resistant weeds in Brazilian agriculture. In: Multifunctional Impacts on Organic and Conservation Agriculture. London: IntechOpen; 2020

[5] Beckie HJ, Ashworth MB, Flower KC. Herbicide resistance management: Recent developments and trends. Plants. 2019:**8**(6):161

[6] Vrbničanin S, Pavlović D, Božić D. Weed resistance to herbicides. In: Herbicide Resistance in Weeds and Crops. London: IntechOpen; 2017

[7] Sherwani SI, Arif IA, Khan HA. Modes of Action of Different Classes of Herbicides. London: IntechOpen; 2015

[8] Varanasi VK, Godar AS, Currie RS, Dille AJ, Thompson CR, Stahlman PW, et al. Field-evolved resistance to four modes of action of herbicides in a single kochia (Kochia scoparia L. Schrad.) population. Pest Management Science. 2015;**71**(9):1207-1212

[9] Tang W, Yu X, Chen J, Xie L, Lu Y. Tolerance to some ACCase inhibitors in four common Roegneria (*Roegneria kamoji*) populations from China. Frontiers in Agronomy. 2020;**2**:14

[10] Takano HK, Ovejero RFL, Belchior GG, Maymone GPL, Dayan FE. ACCase-inhibitingherbicides:Mechanism of action, resistance evolution and stewardship. Science in Agriculture. 2020;**78**(1):2021

[11] Kukorelli G, Reisinger P, Pinke G. ACCase inhibitor herbicides – selectivity, weed resistance and fitness cost: A review. 2013;**59**(3):165-173

[12] Kaundun SS. Resistance to acetyl-CoA carboxylase-inhibiting herbicides. Pest Management Science. 2014;**70**(9):1405-1417

[13] Liu XQ, Yu CY, Dong JG, Hu SW, Xu AX. Acetolactate synthase-inhibiting gametocide amidosulfuron causes chloroplast destruction, tissue autophagy, and elevation of ethylene release in rapeseed. Frontiers in Plant Science. 2017;**8**:1625

[14] Whitcomb CE. An introduction to ALS-inhibiting herbicides. Toxicology and Industrial Health. 2019;**15**(1-2):231-239

[15] Gaines TA, Duke SO, Morran S, Rigon CAG, Tranel PJ. Anita Küpper, et al. Mechanisms of evolved herbicide resistance. The Journal of Biological Chemistry. 2020;**295**(30):10307-10330

[16] Dor E, Galili S, Smirnov E, Hacham Y, Amir R, Hershenhorn J. The effects of herbicides targeting aromatic and branched chain amino acid biosynthesis support the presence of functional pathways in broomrape. Frontiers in Plant Science. 2017;**8**:707

[17] Dong Y, Ng E, Lu J, Fenwick T, Tao Y, Bertain S, et al. Desensitizing plant EPSP synthase to glyphosate: Optimized global sequence context accommodates a glycine-to-alanine change in the active site. The Journal of Biological Chemistry. 2019;**294**(2):716

[18] Busi R, Goggin DE, Heap IM, Horak MJ, Jugulam M, Masters RA, et al. Weed resistance to synthetic auxin herbicides. Pest Management Science. 2018;**74**(10):2265

[19] Todd OE, Figueiredo MRA, Morran S, Soni N, Preston C, Kubeš MF, et al. Synthetic auxin herbicides: Finding the lock and key to weed resistance. Plant Science. 2020;**300**:110631

[20] Dhonukshe P, Grigoriev I, Fischer R, Tominaga M, Robinson DG, Hašek J, et al. Auxin transport inhibitors impair vesicle motility and actin cytoskeleton dynamics in diverse eukaryotes. Proceedings of the National Academy Science USA. 2018;**105**(11):4489-4494

[21] Oettmeier W. Herbicides, Inhibitors of Photosynthesis at Photosystem II. Encycl Agrochem. 2013. Available from: https://onlinelibrary.wiley.com/doi/full/10.1002/047126363X.agr129

[22] Latorre DD et al. Herbicide inhibitors glutamate synthetase. Revista Brasileria de Engenharia de Biossistemas. 2013;**7**(3):134-141

[23] Kraehmer H, Laber B, Rosinger C, Schulz A. Focus on weed control: Herbicides as weed control agents: State of the Art: I. Weed control research and safener technology: The path to modern agriculture. Plant Physiology. 2014;**166**(3):1119

[24] Togo T. Cell membrane disruption stimulates cAMP and Ca2+ signaling to potentiate cell membrane resealing in neighboring cells. Biol Open. 2017;**6**(12):1814-1819

[25] Stewart MP, Langer R, Jensen KF. Intracellular delivery by membrane disruption: Mechanisms, strategies, and concepts. Chemical Reviews. 2018;**118**(16):7409

[26] Wang H, Liu W, Zhao K, Yu H, Zhang J, Wang J. Evaluation of weed control efficacy and crop safety of the new HPPD-inhibiting herbicide-QYR301. Scientific Reports. 2018;**8**(1):3

[27] Tuyen PT, Xuan TD, Thi Tu Anh T, Van TM, Ahmad A, Elzaawely AA, et al. Weed suppressing potential and isolation of potent plant growth inhibitors from Castanea crenata Sieb. Natural Products Chemistry. 2018;**23**(2):345

[28] Dayan FE, Duke SO. Discovery for new herbicide sites of action by quantification of plant primary metabolite and enzyme pools. Engineering. 2020;**6**(5):509-514

[29] Soares da Costa TP, Hall CJ, Panjikar S, Wyllie JA, Christoff RM, Bayat S, et al. Towards novel herbicide modes of action by inhibiting lysine biosynthesis in plants. Elife. 2021;**10**

[30] Umetsu N, Shirai Y. Development of novel pesticides in the 21st century. Journal of Pesticide Science. 2020;**45**(2):54

Chapter 3

New Insight in Herbicides Science: Non-Target Site Resistance and Its Mechanisms

Ermias Misganaw Amare

Abstract

Managing weeds in crop production, whether in the field, or greenhouse, can be troublesome; however, it is essential to successful production. Weeds compete with the crop for nutrients, space, sunlight and also host plant pathogens and insect pests. The economic impacts of weeds include both monetary and non-monetary. In Australia, the overall cost of weeds to grain growers is estimated at AUD 3.3 billion annually. In India, weeds cost over USD 11 billion each year. In the USA, weeds cost USD 33 billion in lost crop production annually. Herbicide use is indispensable in agriculture as it offers tool for weed management; however, repeated applications of herbicides with the same mode of action resulted in the selection of herbicide-resistant weed populations. Herbicide resistance is a rapidly growing worldwide problem that causes significant crop yield losses as well as increases in production costs. Non-target-site resistance to herbicides in weeds can be conferred as a result of the alteration of one or more physiological processes such as reduced herbicide translocation, increased herbicide metabolism, decreased rate of herbicide activation. Non-Target Site Resistance mechanisms are generally more complex and can impart cross-resistance to herbicides with different modes of action. To date, approximately 252 species have evolved resistance to 23 of the 26 known herbicide modes of action.

Keywords: non-target site resistance, absorption, translocation, metabolism

1. Introduction

Weed is one of the main biotic factors that brings about a significant crop yield loss since the beginning of agriculture about 10,000 years ago. Weed will cause the highest potential yield loss to crops. In addition, weeds harbor insects pests, and pathogens, which attack crop plants. Weeds compete with crops for sunlight, water, nutrients, and space. Moreover, weeds infest and destroy native habitats, threatening native plants and grazing lands. Crop yield losses as a result of weeds depend on several factors including weed emergence time, weed density, type of weeds, type of crops, soil fertility, etc. Left uncontrolled, weeds can result in 100% yield loss. In Australia, the overall cost of weeds to grain growers is estimated at AUD 3.3 billion annually. In terms of yield losses, weed loss amounted to 2.7 million tons of grain at a national level [1]. In India, weeds cost over USD 11 billion each year [2]. In India, the yield losses because of weeds were estimated at 36% in peanut, 31% in soybean, 25% in maize, and 19% in wheat. In the USA, weeds cost USD 33 billion in lost crop production annually [2]. Hence, weed management is one of the most important components of cropping systems, which results in significant yield loss as well as increased cost of production. In the early 1950s, synthetic herbicides revolutionized agriculture and have been at the foundation of both weed science research and the intensification and expansion of industrialized agriculture [3].

In developing countries, where farm size is small, weeds management is carried out by hand removal however as a result of rising labor costs and it is being replaced by herbicide use. In most developed countries, herbicides are already widely used to control weeds. However, repeated application of herbicides with similar modes of action has resulted in the development of herbicide-resistant weeds. Currently, more than 500 unique cases of herbicide-resistant weeds have been documented across the globe [4]. The majority of herbicide-resistance weed cases were reported from the USA (more than 160) followed by Australia (over 90 cases) and the remaining cases are reported from Canada, China, and Brazil. The maximum number of herbicide-resistant weed species was reported in different crops, including wheat, maize, rice, soybean spring barley, canola, and cotton [4]. These crops are the most widely produced food crops as well as the important industrial crops. Glyphosate is the most traded herbicide across the globe and used for non-selective post-emergence control of both annual and perennial weeds [3]. This herbicide disrupts the activity of enzymes including 5-enolypyruvyshikimate-3-phospahate synthase [5, 6].

2. Weed management methods

Managing weeds in crop production, whether in the field, greenhouses, or outdoor containers, can be troublesome; however, it is essential to successful production. Weeds not only compete with the crop for plant nutrients, space, and sunlight but also serve as an alternative host to virulent plant pathogens and notorious insect pests. The economic impacts of weeds include both monetary and non-monetary. For example, blackberries restrict human and animal access, harbor pests, reduce pasture production, impede establishment of plants, and reduce naturalness and biodiversity [7]. Some of the common weed management practices are explained below. Weed management activities include preventive/ quarantine, use of cover crops, mowing, flaming, mulching, solarization, and herbicide application.

2.1 Herbicides

In modern agriculture, herbicide spray are very common and rapid weed management method in many crop production areas across the globe. By using herbicides before weeds emerge, weed competition with the crop can be reduced or eliminated, resulting in higher yield and fewer labor costs despite ecological disturbance and health hazards. Herbicides are generally classified according to time of application to the crops and weed growth stage. Preplant herbicides are applied before planting. These herbicides are used before the desirable plants are present because some can control both germinating seedlings and established plants. Pre-emergence herbicides kill weeds at the seed germination stage. These herbicides are applied before weeds emerge. Post-emergence herbicides are applied after the weeds have emerged. Pre-emergence and post-emergence herbicides may be applied before or after the crop is planted depending on the crop and the herbicide selected. However, their extensive utilization across the globe imposes strong selection pressure on resistant weed populations, threatening our ability to successfully manage weed populations. Herbicide resistance is a rapidly growing worldwide problem that causes significant crop yield and quality losses as well as increases production costs. To date, approximately 252 species have evolved resistance to 23 of the 26 known herbicide modes of action, representing over 161 different herbicides [8].

3. Herbicide resistance

The acquired inheritable trait of plants to survive and reproduce under herbicide exposure is defined as resistance. The Weed Science Society of America defines herbicide resistance as the inherited ability of a plant to survive and reproduce following exposure to a dose of herbicide normally lethal to the wild type. Under continuous selection pressure, that is, the repeated use of herbicides with the same mode of action, the resistant weed plants increase in frequency over time, resulting in the domination by individuals resistant to a particular herbicide. Biological and genetic factors of weed species, properties of herbicides, and agronomic practices play a significant role in the evolution and spread of herbicide resistance [8]. Biological characteristics of troublesome weeds, including prolific seed production, high germination percentage, seed dispersal, and longevity, help to maintain a high frequency of resistant individuals in the population. Genetic factors, such as natural mutations conferring herbicide resistance, inheritance of herbicide-resistant genes in the weed population, and fitness of resistance genes in the presence or absence of the herbicide, also play an important role in the evolution and spread of herbicide resistance [8].

Mechanisms of herbicide resistance in weeds can be broadly classified into two categories [8, 9] (i) modifications in the herbicide target enzyme (target-site resistance; TSR) and (ii) mechanisms not involving the target enzyme (non-target-site resistance; NTSR). TSR is typically conferred by single major-effect alleles, whereas NTSR is believed to be conferred by multiple small-effect alleles [9]. The TSR mechanisms largely involve mutation(s) in the target site of action of herbicide, resulting in an insensitive or less-sensitive target protein of the herbicide. In such cases, TSR is primarily determined by monogenic traits. Additionally, TSR can also evolve as a result of the over-expression or amplification of the target gene. TSR mechanisms alter the amino acid sequence and/or expression level of the target enzyme, reducing the herbicide's ability to inhibit the enzyme or requiring a greater herbicide concentration to achieve adequate inhibition [10]. TSR to acetolactate synthase (ALS) inhibitors and acetyl-CoA carboxylase (ACCase) inhibitors, two large classes of herbicides used to control grass weeds, is the most widely documented mechanism of resistance [4]. On the other hand, NTSR mechanisms include all mechanisms that reduce the concentration of active herbicide remaining available to interact with the target site protein, as well as mechanisms that allow the plant to cope with inhibition of the target site [10]. NTSR mechanisms include reduced herbicide uptake/translocation, increased herbicide metabolism, decreased rate of herbicide activation, and/or sequestration [10, 11] (**Figure 1**).

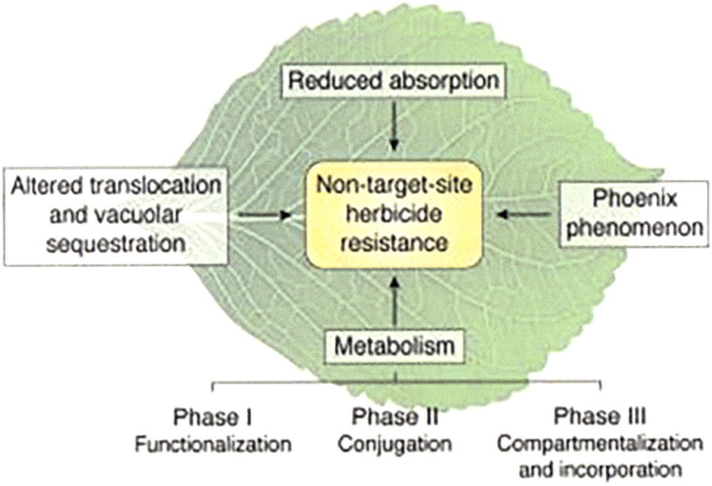

Figure 1.
Weeds can evolve resistance to a herbicide by reducing its absorption, altering translocation and/or sequestration, or developing rapid necrosis of the foliage via degradation of the active ingredient. ***Source:*** *[12].*

3.1 Non-target site resistance (NTSR) mechanisms in weed species

Mechanisms that can contribute to NTSR are complex and involve several different gene types and families. This molecular and genetic complexity makes the identification of particular genes involved in NTSR difficult. Recent advances in this area have identified putative NTSR genes contributing to enhanced herbicide metabolism [13].

3.1.1 Metabolism-based NTSR

Plants contain large numbers of genes encoding enzymes that perform biochemical reactions for the synthesis of secondary metabolites and for detoxifying xenobiotic compounds (e.g., herbicides) [14]. Herbicide metabolism is the degradation of herbicide molecules by endogenous plant enzymes. Metabolism-based NTSR involves increased the activities of enzyme complexes including esterases, cytochrome P450s (CYP450s), glutathione S-transferases (GSTs), and/or Uridine 5′-diphospho (UDP)-glucosyl transferasesm [8]. NTSR, if it involves herbicide detoxification by these enzymes, is usually governed by multiple genes (polygenic) and may confer resistance to herbicides with completely different modes of action [15]. Enhanced rates of herbicide metabolism in NTSR are, in general, have a three-phase process [12, 16].

Phase I reactions increase the polarity of the herbicide and involve oxidation, reduction, or hydrolysis, which form free amino, hydroxyl, or carboxylic acid groups The most common phase I reactions are oxidation reactions carried out by cytochrome P450 monooxygenases (P450s). P450s are a large superfamily of enzymes and catalyze oxygen- and NADPH-dependent monooxygenase reactions [8, 12, 16].

Phase II reactions are commonly catalyzed by the glutathione S-transferase (GST) superfamily that is large and diverse. Higher plants have at least 10 different GST classes, of which the predominant phi and tau classes have broad substrate specificities and are primarily responsible for herbicide detoxification. GSTs conjugate glutathione to oxidized xenobiotics and individual GSTs of several classes are key players in NTSR to herbicides. The best-characterized role of GSTs in NTSR is for the "Peldon" MHR *Alopecurus myosuroides* populations that are resistant to the photosystem II inhibitor chlorotoluron and several ACCase-inhibiting herbicides. The glycosyltransferase enzyme family is also involved in phase II of herbicide metabolism. Specifically, glycosyltransferases conjugate herbicides directly or conjugate a sugar molecule to a variety of lipophilic molecules including xenobiotics. Glycosyl transferases have been shown to metabolize many herbicides and have important roles in conferring tolerance to other abiotic stresses such as salt, cold, and drought, by modifying anthocyanin accumulation [12, 16].

The third phase of herbicide metabolism involves compartmentalization and transportation of the conjugated herbicide into the vacuole or extracellular space [17]. The most common transporters in phase III are ABC transporters. ABC transporters have been shown to transport herbicide metabolites of primisulfuron, glutathione-conjugated herbicide metachlor, and have potential roles in conferring NTSR to glyphosate in *Conyza canadensis* (**Figure 2**).

3.1.1.1 Acetyl CoA carboxylase (ACCase)-inhibitors

Acetyl CoA carboxylase is a very crucial enzyme, which involves in the formation of malonyl CoA *via* the carboxylation of acetyl CoA [8]. Malonyl CoA is needed for *de novo* fatty acid biosynthesis, which is essential for plant survival. ACCase-inhibitors impair malonyl CoA formation in some grass species and ultimately lead to plant death [8]. Research results found metabolic resistance to ACCase-inhibiting herbicides has occurred on many weed plants including Asia minor bluegrass, barnyard grass, blackgrass, Italian ryegrass, Japanese foxtail, rigid ryegrass, and wild oat. In the majority of these cases, enhanced metabolism mediated by CYP450s was reported. For instance, rapid degradation of diclofop-methyl was observed in rigid ryegrass populations from Australia. Interestingly, exposure to low doses of diclofop-methyl acid application is rapidly selected for metabolic resistance in rigid ryegrass. Moreover, the metabolites produced in these resistant plants were found to be similar to those in wheat formed *via* ring hydroxylation and sugar conjugation. This result suggests that in resistant grasses, the metabolism of ACCase-inhibitors occurs through a wheat-like detoxification pathway mediated by CYP450s [8].

3.1.1.2 Acetolactate synthase (ALS)-inhibitors

Enhanced metabolism conferring resistance to ALS inhibitors has been documented in some grass and broadleaf weeds, such as barnyard grass, common waterhemp, Palmer amaranth, rice barnyard grass, rigid brome, short awn foxtail, and water chickweed [8]. Numerous studies have also elucidated the molecular basis of metabolic resistance to ALS inhibitors. Most of the studies have predominantly identified multiple CYP450 genes that are either constitutively expressed or upregulated. For example, the mechanism of mesosulfuron-methyl resistance in short awn foxtail was studied and two CYP450 genes (CYP94A1 and CYP71A4) were overexpressed in the resistant plants. In a similar, two CYP450 genes, (CYP81A12 and CYP81A21) were identified as candidate genes conferring resistance to bensulfuron-methyl and penoxsulam in rice barnyard grass. Several CYP450 genes mediating NTSR to ALS inhibitors have been identified in water chickweed, ryegrass, flaxseed, and blackgrass. In addition to CYP450s, involvement of GSTs, GTs, and ATP-binding cassette (ABC) transporters has also been reported. For instance, in ALS-inhibitor-resistant

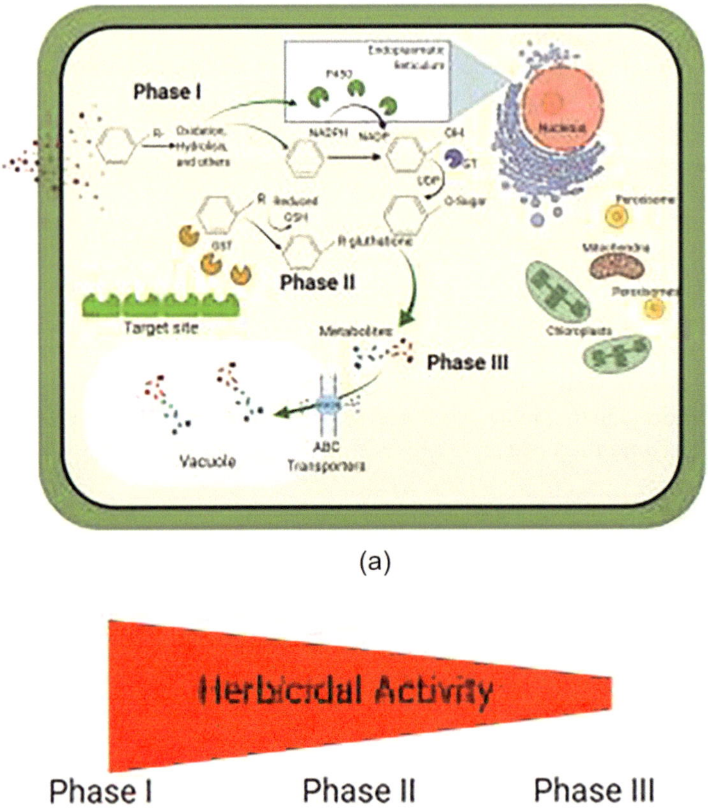

(b)

Figure 2.
Herbicide metabolized in three phases. (a) Initially, herbicide is subjected to a redox reaction to increase its hydrophilicity (phase I). This metabolized herbicide is further processed into phase II. Metabolism may be concluded with the storage of metabolized compounds (phase III). Sources: [15].

water chickweed, four genes including three CYP450s and an ABC transporter were highly expressed in all resistant plants [8].

Recently, a new resistance mechanism in weeds has been identified. Glyphosate resistance is possible through aldo-ketoreductase (AKR)-based metabolism [18], upregulation of an ABC membrane transporter pumping out glyphosate outside the cell [19], and programmed cell death causing rapid necrosis [20]. Similarly, 2,4-D resistance due to either CYP-450-based metabolism [21], a double point mutation [22] or 9-codon deletion in an auxin transcriptional repressor [23], or rapid necrosis [24] has also been reported. These recent findings depicted that herbicide selection for many survival mechanisms will occur and increase the chances for plants to harbor multiple resistance mechanisms.

3.1.1.3 Photosystem II (PSII) inhibitors

The PSII complex is located within the thylakoid membranes of chloroplasts and contains two proteins, D2 and D1 [25]. PS-II inhibitors act by competitively binding to the plastoquinone binding site (QB) on the D1 protein in the PS-II complex of the chloroplast. Once a PSII-inhibiting herbicide binds, it blocks the transfer of electrons from plastoquinone QA in D2 to plastoquinone QB in D1, which prevents CO2 fixation and production of ATP and NADPH. Blocking electron transport leads to the production of reactive oxygen species (ROS), which destroys cell integrity [15, 25]. Some herbicide chemical groups such as triazines, triazinones, and ureas inhibit Photosystem II [15]. Till now, 74 weed species have been reported to develop resistance to PS-II inhibitors across the globe, through both TSR and NTSR mechanisms [25]. NTSR to PS-II inhibitors have been reported in many weed species including bluegrass, common ragweed, common water hemp, Palmer amaranth, and wild radish. The metabolism of PS-II inhibitors was catalyzed by increased activity of GST enzymes and/or CYP450 enzymes [25].

3.1.2 Reduced herbicide absorption

To be effective, herbicides must be absorbed into cells of plants through the roots, in the case of soil-applied herbicides, or from the leaves in the case of foliar-applied herbicides. During herbicide application, herbicide droplets must land on the leaf surfaces and overcome a number of barriers before cellular uptake. This passive process largely depends on leaf surface characteristics, herbicide chemical properties, and their interactions. Herbicide absorption from cellular uptake, where absorption is the process of overcoming the physical barrier of leaves (i.e., cuticle) before the herbicide reaches the apoplast, and uptake is the movement of herbicide from the apoplast into plant cells. Herbicide-resistant weed populations exhibit reduced herbicide absorption, characterized by a reduction in the penetration *via* the cuticle before reaching the epidermis, whereas cell walls do not pose a significant resistance to cellular uptake. Reduced absorption is not a common NTSR mechanism; however, it has occurred in both dicots and monocots to some herbicide groups such as synthetic auxins and 5-enolpyruvylshikimate-3-phosphate synthase inhibitors [15].

Differences in root absorption of herbicides between species have been associated to root morphology differences. There are no cases of evolved resistance to soil-applied herbicides due to reduced root absorption [26]. Differences in foliar absorption of herbicides between weed plants have been highly associated with leaf anatomical structure than biochemical differences [10]. Differential foliar absorption of herbicides between species was directly linked to differences in cuticle thickness and/or composition; however, the number and/or structures of leaf features such as trichomes and hairs have also been involved. For instance, Hirsute leaves are covered with hairy trichomes that can retain spray droplets better than smooth, hairless, or glandless cuticles, hence facilitating absorption. Other leaves have lysigenous glands involved in the production and storage of oily secondary metabolites that can compartmentalize lipophilic herbicides, preventing them from reaching their site of action [26].

Decreased absorption is uncommon NTSR mechanism; however, it has been reported with the resistance of common sunflower to imazethapyr and chlorimuron, prickly lettuce to 2,4-D, annual bluegrass to atrazine, and *L. multiflorum* to glypho sate. No differences were found in cuticular wax amount per unit area of leaf surface between two biotypes of *L. multiflorum* with a threefold difference in glyphosate susceptibility and reduced absorption in the less sensitive biotype. When reduced absorption is implicated, it is most often only one contributing factor to the overall resistance mechanism. For example, resistance to glyphosate in *A. tuberculatus* biotypes was due to both reduced absorption and a herbicide resistance allele of the glyphosate enzyme target EPSPS [12].

3.1.3 Reduced translocation and sequestration

Many foliar-applied systemic herbicides rely on translocation through the phloem. These herbicides must overcome the cuticle barrier and enter the cells of mature source leaves (symplast). This transport can involve active and/or passive diffusion processes [12]. Once inside the symplast, systemic herbicides translocate from source leaves to younger sink leaves *via* the phloem [16]. Herbicide resistance due to reduced translocation occurs when the herbicide is contained in source leaves and prevented from translocating to young leaves. Mechanisms that trap the herbicide in source leaves (e.g., through sequestration within vacuoles of leaf trichomes) or prevent its normal movement to the growing points across membrane barriers (through altered activity of active membrane transporters) will reduce the total amount of herbicide translocated, thus conferring resistance [12]. Therefore, alterations of translocation patterns can lower herbicide efficacy. Herbicide resistance as a result of reduced translocation has been observed in grass weed species, such as *Lolium* spp. [15]. Reduced translocation of glyphosate is the most common type of NTSR mechanism [27]. In these plants, the amount of glyphosate delivered to the meristems is lower than what is essential to be toxic to the weed plant. Reduced glyphosate translocation was first recorded in glyphosate-resistant *Lolium rigidum*, less glyphosate translocated to the meristems, relative to glyphosate-susceptible *L. rigidum* [28]. Glyphosate-resistant *C. canadensis* had reduced translocation [27]. This is due to differences in the cellular distribution of glyphosate and subsequent phloem loading and translocation. In these biotypes, glyphosate enters the source leaves normally; however, it cannot translocate to the meristems because it is rapidly sequestered within the vacuole [29]. Vacuole sequestration activity is temperature-dependent, with less sequestration observed in *C. canadensis* under lower temperatures (**Figure 3**) [30].

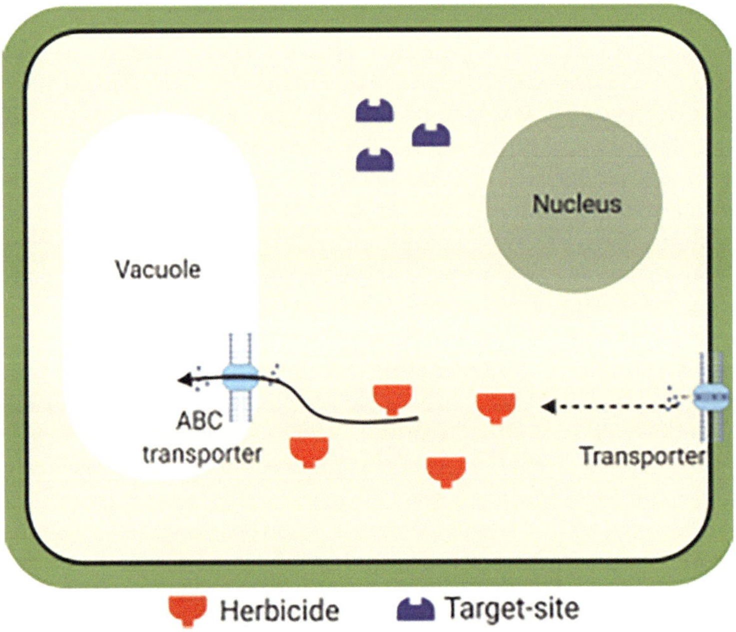

Figure 3.
Reduced herbicide translocation due to vacuolar sequestration. Source: [15].

4. Conclusion

Managing weeds in crop production, whether infield, greenhouses, or containers, can be challenging and costly practice; however, it is essential to successful production. Weeds not only compete with the crop for plant nutrients and sunlight but also host plant pathogens. Herbicides are used in many crop production areas as an economical option to control weeds. By using herbicides before weeds emerge, weed competition with the crop can be reduced or eliminated, resulting in higher quality yield and less labor costs. However, their extensive utilization across the globe imposes strong selection pressure, which results in resistant weed population development. Herbicide resistance is a rapidly growing worldwide problem that causes significant crop yield loss and increases production costs. The most common herbicide resistance form is target-site resistance and non-target site resistance. Non-target site herbicide resistance is complex and involves several different gene types and families. This molecular and genetic complexity makes the identification of particular genes involved in NTSR difficult. Non-target site resistance mechanisms include reduced herbicide uptake/translocation, increased herbicide metabolism, decreased rate of herbicide activation, and/or sequestration. Lack of new herbicides in the market makes utilization of already available herbicides inevitable. Therefore, it is very imperative to integrate various weed management practices to curve a rapid increase in non-target site resistance development. It is equally important to reduce application of the same kind of herbicide over time to overcome resistance to weed population establishment.

Author details

Ermias Misganaw Amare
College of Agriculture and Environment Sciences, Gondar University, Gondar, Ethiopia

*Address all correspondence to: ermiasamare2@gmail.com

References

[1] Llewellyn RS, Ronning D, Ouzman J, Walker S, Mayfield A, Clarke M. Impact of Weeds on Australian Grain Production: The Cost of Weeds to Australian Grain Growers and the Adoption of Weed Management and Tillage Practices. Report for GRDC. Australia: CSIRO; 2016. p. 112

[2] Gharde Y, Singh PK, Dubey RP, Gupta PK. Assessment of yield and economic losses in agriculture due to weeds in India. Crop Protection. 2018;**107**:12-18

[3] Antier C, Kudsk P, Reboud X, Ulber L, Baret P, Messéan A. Glyphosate use in the European agricultural sector and a framework for its further monitoring. Sustainability. 2020;**12**:5682

[4] Heap I. International survey of herbicide-resistant weeds–survey results and criteria to add cases. In: Proceedings of the 16th Australian Weeds Conference. Brisbane: Queensland Weeds Society. Available from: http://www.caws.org.au/awc/2008/awc200810681

[5] Gravena R, Filho RV, Alves PLCA, Mazzafera P, Gravena AR. Glyphosate has low toxicity to citrus plants growing in the field. Canadian Journal of Plant Science. 2012;**92**:119-127

[6] Woodburn AT. Glyphosate: Production, pricing and use world-wide. Pest Management Science. 2000;**2000**(56):309-312

[7] James R, Lockwood M. Economics of blackberries, current data and rapid valuation techniques. Plant Protection Quarterly. 1998;**13**:175-179

[8] Jugulam M, Shyam C. Non-target-site resistance to herbicides: Recent developments. Plants. 2019;**8**:417

[9] Délye C. Unravelling the genetic bases of non-target-site-based resistance (NTSR) to herbicides: A major challenge for weed science in the forthcoming decade. Pest Management Science. 2013;**69**:176-187

[10] Hess FE. Herbicide absorption and translocation and their relationship to plant tolerance and susceptibility. In: Duke SO, editor. Weed Physiology. Boca Raton, FL: CRC Press, Inc.; 1985. pp. 191-214

[11] Devine MD, Eberlein CV. Physiological, biochemical and molecular aspects of herbicide resistance based on altered target sites. In: Roe RM, Burton JD, Kuhr RJ, editors. Herbicide Activity: Toxicology, Biochemistry and Molecular Biology. Amsterdam, The Netherlands: IOS; 1997. pp. 159-185

[12] Gaines TA, Duke SO, Morran S, Rigon CAG, Tranel PJ, Küpper A, et al. Mechanisms of evolved herbicide resistance. Journal of Biochemistry. 2020;**295**:10307-10330

[13] Torra J, Osuna MD, Merotto A, Vila-Aiub M. Multiple herbicide-resistant weeds and non-target site resistance mechanisms: A global challenge for food production. Frontiers in Plant Science. 2021;**12**:763212

[14] Yuan JS, Tranel PJ, Stewart CN. Non-target-site herbicide resistance: A family business. Trends in Plant Science. 2007;**12**:6-13

[15] Suzukawa AK, Bobadilla LK, Mallory-Smith C, Brunharo CACG. Non-target site resistance in *Lolium* spp. globally: A review. Frontiers in Plant Science. 2021;**11**:2137

[16] Burns EE. Genetic and Physiological Characterization and Ecological Management of non-target Site Resistance in Multiple Herbicide Resistant *Avena fatua L.* a Dissertation Submitted in Partial Fulfilment of the Requirements for the Degree of Doctor of Philosophy in Plant Sciences. Bozeman, Montana: Montana State University; 2017

[17] Hu J, Tranel PJ, Stewart CN Jr, Yuan JS, Stewart C Jr. Molecular and genomic mechanisms of non-target site herbicide resistance. In: Ames IA, editor. Genomics of Weedy and Invasive Plants. Ames, IA: Blackwell Scientific; 2009. pp. 149-161

[18] Pan L, Yu Q, Han H, Mao L, Nyporko A, Fan L, et al. Aldoketo reductase metabolizes glyphosate and confers glyphosate resistance in *Echinochloa colona*. Plant Physiology. 2019;**19**:979

[19] Pan L, Yu Q, Wang J, Han H, Mao L, Nyporko A, et al. An ABCC type transporter endowing glyphosate resistance in plants. Proceedings of the National Academy of Sciences of the United States of America. 2021;**118**:e2100136118

[20] Van Horn CR, Moretti ML, Robertson RR, Segobye K, Weller SC, Young BG, et al. Glyphosate resistance in Ambrosia trifida: Part 1. Novel rapid cell death response to glyphosate. Pest Management Science. 2018;**74**:1071-1078

[21] Giacomini DA, Patterson EL, Küpper A, Beffa R, Gaines TA, Tranel PJ. Coexpression clusters and allele-specific expression in metabolism-based herbicide resistance. Genome Biology and Evolution. 2020;**12**:2267-2278

[22] Leclere S, Wu C, Westra P, Sammons RD. Cross-resistance to dicamba, 2, 4-D, and fluroxypyr in Kochia scoparia is endowed by a mutation in an AUX/IAA gene. Proceedings of the National Academy of Sciences. 2018;**115**:2911-2920

[23] Figueiredo MR, Kuepper A, Malone JM, Petrovic T, Figueiredo ABT, Campagnola G, et al. An in-frame deletion mutation in the degron tail of auxin co-receptor IAA2 confers resistance to the herbicide 2, 4-D in Sisymbrium orientale. bioRxiv. 2021

[24] De Queiroz AR, Delatorre CA, Lucio FR, Rossi CV, Zobiole LH, Merotto A. Rapid necrosis: A novel plant resistance mechanism to 2, 4-D. Weed Science. 2020;**68**:6-18

[25] Gronwald J. Resistance to photosystem II inhibiting herbicides. In: Herbicide Resistance in Plants: Biology and Biochemistry. Boca Raton, FL, USA: Lewis Publishers; 1994. pp. 27-60

[26] Devine M, Duke SO, Fedtke C. Physiology of Herbicide Action. Englewood Cliffs, NJ: Prentice Hall; 1992

[27] Gaines TA, Patterson EL, Neve P. Molecular mechanisms of adaptive evolution revealed by global selection for glyphosate resistance. The New Phytologist. 2019;**223**:1770-1775

[28] Lorraine-Colwill DF, Powles SB, Hawkes TR, Hollinshead PH, Warner SAJ, Preston C. Investigations into the mechanism of glyphosate resistance in Lolium rigidum. Pesticide Biochemistry and Physiology. 2002;**74**:62-72

[29] Ge X, d'Avignon DA, Ackerman JJH, Sammons RD. Rapid vacuolar sequestration: The horseweed glyphosate resistance mechanism. Pest Management Science. 2010;**66**:345-348

[30] Ge X, d'Avignon DA, Ackerman JJH, Duncan B, Spaur MB, Sammons RD. Glyphosate-resistant horseweed made sensitive to glyphosate: Low-temperature suppression of glyphosate vacuolar sequestration revealed by P-31 NMR. Pest Management Science. 2011;**67**:1215-1221

Section 2

Mixture, Management and Environmental Impact

Chapter 4

Joint Action of Herbicides on Weeds and Their Risk Assessment on Earthworm (*Eisenia fetida* L.)

Mohammad Taghi Alebrahim, Elham Samadi Kalkhoran and Te-Ming Paul Tseng

Abstract

Frequent and intensive use of similar modes of action herbicides increases selection pressure resulting in nature adapt and a number of herbicide-resistant weeds. The most effective methods to prevent and delay herbicide-resistant weeds are herbicide tank mixture and adjuvant mixed herbicides. This chapter intends to explain the advantages of herbicide tank mixture and adjuvant mixed herbicides. In addition, the models of estimated herbicide mixture interaction response have been explained. Although herbicide mixtures have benefits, they may present risks leading to soil pollution and affecting soil fauna such as earthworms. Therefore, we discussed the negative effect of mixture herbicides on *Eisenia fetida*. On the other hand, various models to calculate mixture herbicide toxicity on earthworms will be present in this chapter.

Keywords: adjuvant, chemical control, earthworm, estimated model herbicide mixture

1. Introduction

Heavy reliance on herbicides has increasingly raised environmental concerns [1–3]. The selection pressure of herbicides resulted in nature adapting and eventually developing herbicide-resistant and tolerant weeds biotype [4–7]. The most effective tool to inhibit, delay, or control herbicide-resistant weeds is to substitute herbicides with different modes of action [8, 9]. But numerous studies have been conducted that simple switches do not delay the evolution of resistant weeds [10, 11]. Previous studies have shown that combining multiple herbicide modes of action in tank mixtures is more efficient in managing weeds [10, 12]. Mixing various modes of action in the mixture can control resistant weeds via broadening the selection pressure by targeting multiple metabolic pathways and delaying the evolution of herbicide-resistant weeds [13]. Ideal herbicide mixtures have proven beneficial over using a single herbicide in improving control and broadening the weed control spectrum [14, 15]. It contains active components with the same persistence and spectrum of controlled weeds but through a different mode of action [16]. Tank mixing increases in a spectrum of controlled weeds or an extension of weed control over a more extended period, which reduces production cost by saving time and labor, reduces the number of machine entrances into the production area, fuel consumption, water use to prepare the solution, and hours spent. This leads to lower soil compaction by eliminating multiple field operations. Crop safety is improved by adopting a combination of selected herbicides with minimum doses rather than a single high amount of one herbicide. The soil residues of persistent herbicides were decreased following the application of the minimum levels of such herbicides [17]. It is presupposed that herbicide tank mixtures with two or more herbicide partners behave and act independently so that the presence of each one does not affect the activity of another or may significantly modify the biological behavior of every herbicide in the mixture. Regarding the herbicide tank mixtures, the activity of the applied combination can be easily predicted as the sum of the activities related to each herbicide when applied separately.

In some cases, the interactions often result in declining or enhancing the activity of the combined herbicides compared with the sum. Practically, the herbicide combinations exhibit more activity on target weed species and less on crops (higher selectivity). However, the prediction of this issue is difficult since the behavior of each herbicide in the mixture is mainly influenced by the presence of the other(s), and the mixture activity may significantly vary depending on plant species, growth stage, and environmental conditions. Multiple herbicides applied in the mixture have three types of herbicide interaction: additive/neutral, synergistic, or antagonistic [18–20] (**Figure 1**). Synergism is favorable when two or more herbicide mixtures perform rather than the herbicides applied alone. It allows a lower application rate or frequency of herbicide treatment [22], but finding a new synergy remains challenging. In contrast, an antagonistic response is an interaction of two or more herbicides such that the effect, when combined, is less than the predicted effect based on the activity of each chemical applied separately. Antagonism is 2–3 times more common than synergy, especially when herbicides from different chemical families are combined [21]. Sometimes, synergism can be hypothesized based on mechanistic assumptions, as was done by [23], who predicted the synergism between glufosinate and protoporphyrinogen oxidase inhibitors and confirmed it experimentally; but generally,

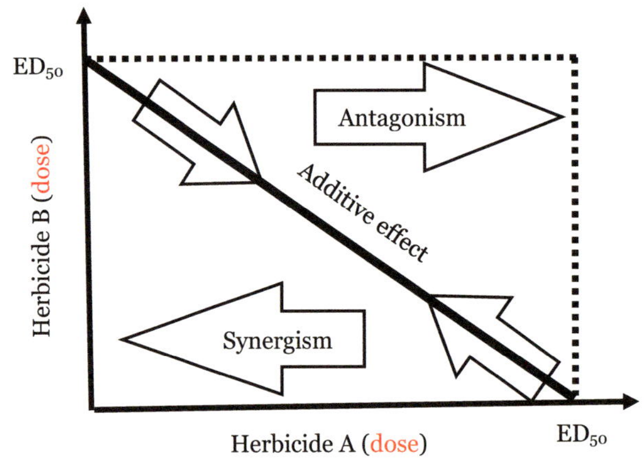

Figure 1.
Schematic isobologram for additive, synergism, and antagonism response of herbicide interaction (ED_{50} = herbicides doses, applied singly or in the mixture for 50% weed control) (modified from [21]).

synergies are not predictable. A synergistic herbicide mixture for one species can also be antagonistic or additive for another species [24]. Thus, herbicide synergies appear to be rare and unpredictable. An additive/neutral response occurs when the observed response of two jointly applied herbicides is statistically similar to the expected value of the mixture. The interactions in herbicide mixtures can occur before, during, or after utilizing the mixture, the mechanisms of which can be broadly grouped into biochemical, competitive, physiological, and chemical categories [25]. This chapter aims to explain the importance of herbicide mixtures for weed control and to clarify the models to estimate combined herbicides' effects. Meanwhile, discusses the risk assessment of herbicide mixtures on the earthworm population.

1.1 Models used to estimate mixture herbicide interaction

The use of isobologram could determine the synergism and antagonism response of the mixtures [26]. Isobologram is a two-dimensional graph. There are two dose axes, x and y, in the mixtures. Herbicide A is the dose on the x-axis, and herbicide B is the dose on the y-axis. The mixtures follow the additive response when mixtures do not interact and present straight lines, and the analysis of this mixture is based on the additive dose model (ADM) [27]. The mixtures may interact, and the performance of combined herbicides is greater than that of herbicides applied alone. So, herbicides are more effective than expected and followed synergism. It means using a lower dose of combined herbicides to provide the same effect as herbicides applied alone. In contrast, if the efficacy of the herbicide mixture is less than that applied alone, then they show antagonism [26].

The reference model uses to determine synergism, antagonism, and additive response in the mixtures. Any consistent model must relate biological response to the doses of two or more herbicides. Choice of the reference model is crucial as the different models may produce different conclusions. The two most frequently referenced models in the study of joint action will be referred to as the additive dose model (ADM) and the multiplicative survival model (MSM) [28]. ADM assumes additivity of doses, i.e., that one herbicide can be replaced, wholly or partly, by another herbicide at equivalent doses. In contrast, MSM assumes that the expected efficacy of herbicide mixtures can be calculated by multiplying the percent survivals of the individual herbicides. Hence, a fundamental difference between the two models is that ADM considers dose rates, whereas MSM considers effects. Both dose addition and independent action should be helpful to approximations for defining the predicted response in the absence of herbicide interactions. A widely known characteristic of the ADM is that, for mixtures of two components, when the response surface predicted by the model is plotted against arithmetic scales of the component doses, the contours of equal response (i.e., isobols) are straight lines. At any particular level of response, the relative potency of the components when acting alone establishes scales of equivalent doses. In terms of this effective-dose (ED) scale, if one component of the mixture is replaced, wholly or in part, by the other, the predicted response is unchanged. By contrast, the MSM does not generally give straight-line isobols. The distinction between the ADM and MSM has not consistently been recognized, and different analysis methods have been confused with other models.

A third reference effect, effect addition, has been proposed, although it predicts implausible effects under certain realistic conditions [29, 30]. Therefore, it is unlikely to be helpful in practice. Likewise, the evaluation of adjuvants does not elicit any antagonistic or synergistic effects since there is no comparison with a reference effect, and it is the only so-called enhancement or potentiation effect [30]. There are various types of herbicide mixtures, experimental designs, and used models. A single-dose factorial design and multiple-dose factorial design are two main groups.

1.1.1 A single-dose factorial design

Two factors are involved in fixed-dose or single-dose experimental design. The first factor is several herbicides (two herbicides), and the second factor is dose with two levels (dose 0 and a nonzero dose). Overall, four treatments result in this design: control (dose 0 of both A and B) (E_O), a nonzero dose of A and dose 0 of B (E_A), dose 0 of A and a nonzero dose of B (E_B), and a single mixture dose corresponding to nonzero doses of both A and B (E_{AB}) [31].

Two nonzero doses justify certain model assumptions despite playing no role in the subsequent derivation. Thus, the doses should be carefully selected since any claim about an antagonistic or synergistic effect is only valid for the chosen doses. Synergism or antagonism can influence dose selection so that the use of a full recommended dose of each pesticide may mask potential synergism when trying to detect synergism for two highly effective herbicides. In this case, pesticide dose reduction (e.g., by 50%) is a common solution. The statistical analysis of 2×2 factorial design is based on the ordinary or linear mixed two-way Analysis of Variance (ANOVA) model depending on the experimental design [32]. It is assumed that fitting the two-way ANOVA model leads to the four estimates of E_O, E_A, E_B, and E_{AB}. In this regard, the subscript 0 refers to the control, A and B are considered as the separate effects of A and B, respectively, and AB indicates their combined effect. Regarding the ordinary two-way ANOVA, the estimates are simple treatment means for each group, while the weighted mean for the linear mixed one. Comparing E_O, E_A, E_B, and E_{AB} through pairwise comparisons does not demonstrate any antagonistic or synergistic effects after fitting a two-way ANOVA model. An antagonistic or synergistic effect may be reported where there is none. Further, the estimates can be used to derive the predicted effect under the assumptions of dose addition and independent action.

1.1.1.1 Dose addition

The reference effect (E_{add}) under the assumption of dose addition is defined as follows [33]:

$$E_{add} = (E_A - E_O) + (E_B - E_O) \tag{1}$$

The definition in Eq. (1) may be justified as reflecting dose addition (even though effects and not doses are added up) by supposing linear dose-response relationships for the two pesticides [32]. Given the availability of only a single nonzero dose for the two pesticides, it is not meant to assume any nonlinear dose-response relationships. However, a linear dose-response relationship may often be assumed as a local approximation to the true nonlinear relationship. This assumption can be justifiable if amounts were chosen as the effective doses, which are not too extreme since the dose-response relationship within a restricted dose range may be supposed to be approximately linear. Particularly, let $y_A = a_0 + b_A x_A$ and $y_B = a_0 + b_B x_B$ denote the simple linear regression equations for the two pesticides with the response values of y_A and y_B, as well as the doses of x_A and x_B, respectively. Then, the reference effect E_{add} is as follows:

$$E_{add} = (a_0 + b_A x_A - a_0) + (a_0 + b_B x_B - a_0) = b_A x_A + b_B x_B \quad (2)$$

representing that the sum of effects is equal to that of doses after appropriate scaling [34]. Each antagonistic or synergistic effect can be defined as the difference (D_{DA}) between the observed response (expressed as the difference from the control) and predicted effect (Eq. (1)). Especially, the difference is considered as follows:

$$D_{DA} = E_{AB} - E_0 - E_{add} = E_{AB} - E_0 - (E_A - E_0 + E_B - E_0) = E_{AB} - E_A - E_B + E_0 \quad (3)$$

Based on the definition of difference D_{DA} in Eq. (3), the values significantly larger and smaller than zero exhibit a synergistic and an antagonistic effect, respectively. Testing the null hypothesis of no antagonistic or synergistic effect corresponds to testing for no interaction in a standard two-way ANOVA model. Regarding reporting, the difference must be accompanied by the corresponding standard error or 95% confidence interval to allow for the uncertainty attached to the estimate.

1.1.1.2 Independent action

The reference effect (E_{ind}) under the assumption of independent action is defined as follows:

$$E_{ind} = E_0 \left(1 - \frac{E_0 - E_A}{E_0}\right)\left(1 - \frac{E_0 - E_B}{E_0}\right) = E_0 \left(\frac{E_A.E_B}{E_0.E_0}\right) = \frac{E_A.E_B}{E_0} \quad (4)$$

as rephrasing in terms of the parameters in the two-way ANOVA model [35]. Similar to the dose addition, the reference effect only involves the three estimates corresponding to the control group (E_0) and the two separate effects of pesticides A and B (E_A and E_B, respectively). In contrast to the definition of dose addition in Eq. (4), which only includes contrasts (i.e., the differences relative to the control), the definition in Eq. (3) relies heavily on the absolute level of the control group (E_0). Furthermore, any antagonistic or synergistic effect may be expressed as the discrepancy between the observed and reference effect under the assumption of independent action in the same way as for dose addition. The difference (D_{IA}) is defined as follows:

$$D_{IA} = E_{AB} - E_{ind} = E_{AB} - \left(\frac{E_A.E_B}{E_0}\right) \quad (5)$$

The difference D_{IA} significantly below or above zero demonstrates an antagonistic or synergistic effect, respectively. The difference should be reported with the corresponding standard error or 95% confidence interval, which can be obtained by using the delta method. The delta approach is a statistical technique for estimating the standard errors of derived parameter estimates (i.e., the parameters that do not explicitly feature the model parameterization) [18].

1.1.2 Multidose factorial designs

The multidose design is similar to the single-dose one except that a dose range is selected for one or both pesticides, and mixture doses are obtained based on a complete or incomplete two-way factorial design (**Figure 2**). The statistical modeling

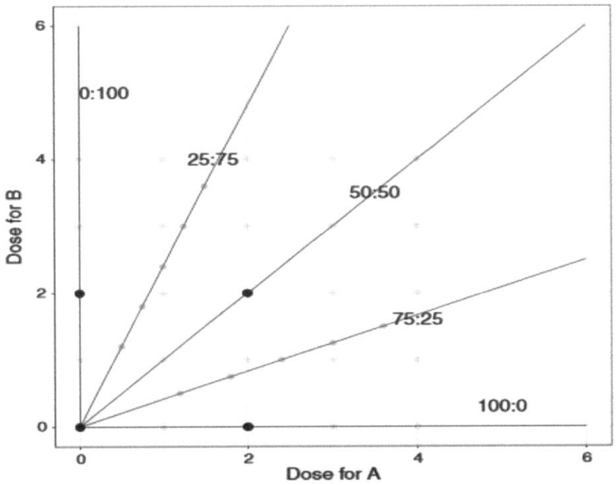

Figure 2.
Factorial and fixed-ratio designs for binary mixture experiments (black and light-gray points illustrate fixed-/ single-dose and multidose factorial designs, respectively. The dark-gray lines reflect the rays in a fixed-ratio design with five rays. In addition, three mixtures (virtual proportions of 25:75, 50:50, and 75:25) and two degenerate mixture rays are observed for the individual pure pesticides (virtual proportions of 100:0 and 0:100). The dark-gray points represent the amounts selected along the rays. The doses for the factorial and fixed-ratio designs hardly overlap [33].

approach outlined for single-dose designs can be simply applied in multidose designs by analyzing one mixture dose at a time in the separate statistical analyses corresponding to fitting two-way ANOVA models. A multidose design can be considered as a collection of single-dose designs, and a design involving multiple mixtures in single doses can be analyzed in the same way. However, this method may or may not imply the suboptimal use of data depending on the type of response and experimental design. Fitting a simultaneous model and borrowing strength across mixture doses may improve the analysis in some cases [36].

1.1.3 Single-ray fixed-ratio designs

The single-ray mixture fixed-ratio design consists of several mixture doses so that the two individual herbicides contribute to doses in a constant ratio (in a single ray), which may be specified in terms of so-called actual or virtual proportions. Further, the design involves the two rays corresponding to the individual, pure pesticides, utilized in several doses. Determining total mixture doses is an important preliminary step in planning a fixed-ratio mixture experiment. These doses can be used for subsequent dose-response modeling. Ideally, this step requires prior knowledge about effective doses. Therefore, it is assumed that ED_{50A} and ED_{50B} are available from the previous experiment. The resulting relative potency of pesticide B relative to A is denoted ρ ($=ED_{50B}/ED_{50A}$). For a given mixture fraction $f \in [0, 1]$, which is respectively related to virtual (mixture) proportions f and $1-f$, the corresponding actual mixture proportions f_A and f_B (the relative potency of the pesticides A and B) can be calculated as:

$$f_A = f ED_{50A}/(f\ ED_{50A} + (1-f)(ED_{50B}) \qquad (6)$$

$f_B = 1 - f_A$. This approach for extracting the actual mixture proportions is referred to as Hewlett's criterion, which is optimal compared with the other methods [31]. For instance, if the ED_{50} values of herbicides metribuzin and flumioxazin are respectively equal to 17 and 153 μg cm^{-2} in a preliminary experiment, then, a virtual 50:50 mixture (f = 0.50) corresponds to an actual 10:90 mixture by using Eq. (6) (with the actual mixture proportions of 0.10 and 0.90 for metribuzin and flumioxazin, respectively). The ED_{50} value under the assumption of dose addition, ED_{50add} (expressed as a total dose), can be obtained by using either actual or virtual mixture proportions as $ED_{50A}/(f_A + f\rho) = fED_{50A} + (1 - f)\ ED_{50B}$ [34]. Based on the actual proportions f_A and f_B, the doses of A and B in the mixture can be respectively recovered as $f_A ED_{50add}$ and $f_B ED_{50add}$ (they are needed for the practical application of the mixture). The resulting ED_{50add} and corresponding doses A and B are typically used to derive a dose series through repeated twofold decreases and increases [37]. The number of doses should be guided by the same considerations utilized for the ordinary dose-response curves of single pesticides. Additionally, no preliminary experiments are carried out in some cases. As an approximation, the relative potency can be estimated from the dose-response data for pesticides A and B, obtained as a part of the ongoing mixture experiment. However, it should be noted that the resulting doses for the mixture are partly based on the estimates (which are based on the response data). The uncertainty in these estimates is ignored in a standard statistical analysis. The data of three dose-response curves can be used to assess synergistic and antagonistic effects on the dose scale [38]. The presence of a shared control group (for dose 0) in dose-response curves is an important prerequisite. This assumption is usually ensured by the experimental design. It implies an indirect standardization relative to the control, which is not unlike the use of differences relative to the control in the case of factorial designs. A joint dose-response model should be fitted for continuous response data, while dose-response models may be separately fitted for each ray concerning binomial and count response data.

1.1.3.1 Dose addition for fixed-ratio designs

Three scenarios are distinguished depending on how similar or dissimilar the dose-response curves are assumed. The assumptions have profound implications on how to evaluate antagonistic and synergistic effects.

1.1.3.2 Identical lower limits and slopes: dose-response models

That imposing shared lower and upper limits and slopes for all three dose-response curves often referred to as parallelism have been used for a long time. These models involve only a single parameter for the common lower and upper limits, slope, and three parameters for the ED_{50} (one for each curve). Accordingly, there are a total of six model parameters. Under the assumption of dose addition, the estimated mixture ED_{50} (ED_{50add}) can be calculated by the linear combination of the ED_{50} values estimated for individual pesticides as [33]:

$$ED_{50add} = f\ ED_{50A} + (1 - f)ED_{50B} \tag{7}$$

by using the virtual proportions f and $1 - f$ [39]. It is important to realize that ED_{50add} is a derived estimate and consequently is determined with uncertainty like other estimates. Further, Eq. (7) is equivalent to the commonly shown but less intuitive equation for dose addition in terms of so-called toxic units [33]:

$$\frac{f_A ED_{50add}}{ED_{50A}} + \frac{f_B ED_{50add}}{ED_{50B}} = \frac{x_A}{ED_{50A}} + \frac{x_B}{ED_{50B}} = 1 \quad (8)$$

where $x_A = f_A ED_{50add}$ and $x_B = f_B ED_{50add}$ are respectively considered as the total doses of pesticides A and B in proportions f_A and f_B, leading to an effect corresponding to $ED50_{add}$. In the following, Eq. (8) is only utilized because of offering a much more direct interpretation of dose addition [39].

Fitting the dose-response model(s) results in estimating ED_{50A}, ED_{50B}, and ED_{50mix} (expressed as total doses). Furthermore, both a difference and a ratio may be used to examine departures from the assumption of dose addition. In any case, the corresponding standard error or 95% confidence interval should be reported, the first of which can be computed by employing the delta method. Particularly, the definition of the difference is as follows [33]:

$$D_{DA} = ED_{50mix} - ED_{50add} \quad (9)$$

An estimated difference significantly more or less than zero reflects an antagonistic or synergistic effect. It is worth noting that ED_{50add} and ED_{50mix}, which do not incorporate the uncertainty of both estimates, should not be compared [40]. The ratio, combination, or interaction index is defined as follows [32]:

$$R_{DA} = \frac{ED_{50mix}}{ED_{50add}} \quad (10)$$

where a value significantly larger than 1 illustrates an antagonistic effect, while a synergistic effect is detected when a value is significantly lower than 1. The use of arbitrary cutoffs such as $R_{DA} < 0.8$ and > 1.2 is not enough for declaring synergism or antagonism, respectively, since the variation in R_{DA} is ignored entirely. The utilization of a difference in terms of logarithm-transformed estimated ED_{50} values corresponds to the application of ratio R_{DA}. These difference and ratio respectively expressed by Eqs. (9) and (10) need not lead to the same results because of using various approximations while calculating the corresponding standard errors based on the delta approach.

1.1.3.3 Identical lower limits but varying slopes

In log-logistic and Weibull dose-response models, the approximations of estimates for the slope parameter b and parameter e (ED_{50} in the log-logistic one) have recently been established by supposing dose addition [41]. The approximations can be compared with the parameters estimated for the fitted dose-response curve of the mixture. Regarding the log-logistic model, this approach provides a framework for comparing the observed ED_{50} for the mixture with the predicted ED_{50} under this assumption. The approximation of ED_{50} coincides with Eq. (7) for the identical slope scenario. In addition, a slight difference is observed in the approximations for the identical and varying slope scenarios [42]. Thus, varying slopes may not warrant a different analysis than for the earlier case of identical slopes and lower limits when interest lies in ED_{50}. In other words, Eqs. (7), (9), and (10) may still be applied for assessing synergistic and antagonistic effects. However, a different definition of reference effect under the assumption of dose addition may be required for varying slope scenario if interest is in other effective doses [42].

1.1.3.4 Varying slopes and varying lower limits

The varying lower limits may be caused by the lack of absorption or solubility, complicating the evaluation of synergistic and antagonistic effects. For example, the assumption of dose addition needs to no longer correspond to the linear relationships between effective doses (Eq. (7)) [43]. A crude approximation is obtained by supposing identical limits, which should be flagged during use. The literature has proposed several approaches for handling varying lower limits or relevant varying upper limit scenario. Further, many generalizations of existing dose-response models have been suggested [44], often involving highly nonlinear regression models or additional assumptions to present suitable predictions. However, the generalizations are not yet readily available to practitioners. The estimation and quantification of departure from the reference effect remain difficult. The utilization of an absolute effect level, which is separately reached for both pesticides, can be addressed as an alternative. The corresponding (relative) effective doses need not correspond to (relative) ED_{50}, although they are defined independently of the lower limit (as if the lower limit is zero for both pesticides). This approach can provide a viable solution in pesticide science since the control (dose 0) mostly corresponds to the highest response level. Differing lower limits often occur for relatively high doses. The procedure previously described for the case with identical slopes and lower limits can be employed in the case of selecting the appropriate absolute effect level. However, the definition of the effective dose under the assumption of dose addition may not be straightforward for the varying slope scenario.

1.1.4 Independent action for fixed-ratio designs

In analogy with Eq. (4), the dose-response function for the mixture f_{ind} under the assumption of independent action is defined from the dose-response functions f_A and f_B for individual pesticides as f_{ind}:

$$f_{ind}(x) = \frac{f_{A(x)} \cdot f_{B(x)}}{f_{A(0)}} \tag{11}$$

for any dose x. The denominator can be the mean response level at dose zero for each of the two individual pesticides, which should have the same upper limit by the assumption. In many applications, in which the response values are pre-standardized against the control [45], Eq. (11) reduces to simply being the product (e.g., standardization means $f_A(0) = f_B(0) = 1$ in Eq. (11) f_{ind}:

$$f_{ind}(x) = f_A(x) \cdot f_B(x) \tag{12}$$

With respect to mathematical form, the function f_{ind} expressed by Eqs. (11) or Eqs. (12) is not the same as the model functions f_A and f_B for individual pesticides. Accordingly, log-logistic models for individual pesticides do not imply a log-logistic model under the assumption of independent action. However, the upper limits of function f_{ind} and two individual functions are identical [41]. Furthermore, the lower limit of find equals zero if one of the model functions f_A and f_B has a lower limit of zero. The entire estimated dose-response curve for the mixture. The entire estimated dose-response curve for the mixture can be compared with the predicted dose-response curve under the assumption of independent action obtained from Eqs. (11) or Eqs. (12)

through visual inspection or statistical tests such as two-sample t-tests or nonparametric equivalents (comparing fitted and predicted values dose by dose) [46]. The statistical methods suppose the independence between fitted and predicted values, so they are not entirely appropriate. In other words, the assumption of independent action is amenable for predicting, not for quantifying antagonistic or synergistic effects in terms of mean departures from the reference effect in the fixed-ratio ray design.

1.1.5 Multi ray fixed-ratio designs

In the case of an experimental design with multiple mixture rays (**Figure 2**), the earlier methods for the identical and varying slope scenarios for ED50 may still be implemented, repeating the analysis for each mixture ray. Since these separate analyses share the same control group, some overlaps are detected in the used data, although they may be acceptable [47].

1.2 Review of research on the effects of herbicides mixtures on weeds

We note in this section several research results that concluded additivity, antagonism, and synergism effects on weeds.

One of the most common herbicide mixtures is different graminicides with broadleaf herbicides mixture to broaden the weed control spectrum. The postemergence application of various graminicides in a mixture with one or more broadleaf herbicides often results in reduced efficacy of graminicides [48]. Antagonistic interactions are probably due to morphological and physiological differences between grasses and broadleaf weeds. Broadleaf weeds have meristems at the top of the plant, whereas grasses have them at the base. On the other hand, this difference affects absorption and mainly translocation of the foliar-applied herbicides, particularly the systemic ones that are translocated and accumulated at the meristematic tissues of the plant where they act. The herbicide amount translocated to its site of action can be declined by the presence or concomitant translocation of another herbicide into the plant [48]. Increasing the ratio of graminicide to broadleaf herbicide in a mixture can alleviate the antagonism of the graminicide [49]. Historically, ACCase inhibiting herbicide antagonism has been observed when applied in a mixture with broadleaf or sedge herbicides, such as ALS inhibiting herbicides and photosystem II inhibiting herbicides [19, 50]. Research by [19] showed that quizalofop (120 g ha^{-1}) mixed with the full labeled rate of halosulfuron at 53 g ha^{-1} could result in an antagonistic interaction for weedy rice and barnyardgrass control. The interaction of herbicides in-tank mixing depended on weed species. Noticeably, the highest dose of halosulfuron (53 g ha^{-1}) mixed with quizalofop followed an additive response on red rice (*Oryza punctata*) 28 days after treatment [51, 52]. Glufosinate antagonized the activity of clethodim on a mixed population of annual grass species: large crabgrass and fall panicum (*Panicum dichotomiflorum* Michx.), goosegrass (*Eleusine indica* L.) [53], and giant foxtail (*Setaria faberi* Herrm.) [51]. However, [54] did not identify antagonism of glufosinate + clethodim on barnyardgrass. Weed's different responses to herbicide interactions may be due to genetic, physiological, or morphological differences [25]. Antagonism of an ACCase inhibiting herbicide can be reduced by increasing the rate of the ACCase inhibitor to broadleaf herbicide in a mixture. The antagonism between bentazon and quizalofop for control of barnyardgrass (*Echinochloa crus-galli*) can be overcome by doubling the rate of quizalofop [55]. Antagonistic interactions may be attributed to the increased metabolism of an herbicide in the presence of another. Based on the study results [56], the less efficacy of diclofop on various species following application with hormone herbicides such as 2,4-D is ascribed to an enhancement in its metabolism (complex formation) carboxylic group)) due to the presence of 2,4-D. The previous studies revealed that the members of aryloxyphenoxypropionate and cyclohexanedione herbicides are more affected when mixing with systemic broadleaf herbicides than the contact ones. The interaction of herbicide mixtures depends on dose and growth stages. Glufosinate at 451 g ha^{-1} + clethodim at 76 g ha^{-1}, an improvement in control was observed over the individual herbicides for barnyardgrass and Johnson grass (*Sorghum halepense*) control. In contrast, a reduction was observed for large crabgrass (*Digitaria sanguinalis*) and no difference for broadleaf signalgrass [57]. Additionally, the extent of the interactions between combined herbicides is mostly influenced by the growth stage of weeds. The post-emergence use of chlorsulfuron and diclofop diminishes the efficacy of diclofop on Italian ryegrass (*Lolium multiflorum*), the effect of which is more severe when the application is performed at the three-leaf growth stage than the two-leaf one [58]. This issue may be related to a reduction in detoxification ability compared with the younger plants, as well as their thinner cuticle, which probably allows to retain, absorb, and translocate the greater amounts of the utilized herbicides. In the research of [59], the antagonism effect was observed when 28.5% nicosulfuron mixed mesotrione by ADM model on canola at 10, 17, and 40 days after treatment. An increased level of Reactive Oxygen Species (ROS), produced by the mesotrione, may block the inhibitory effect of nicosulfuron on ALS [55]. Clomazone at 760 g ha^{-1} + 1540 g ha^{-1} pendimethalin mixed with 1120 or 2240 g ha^{-1} propanil followed an antagonistic effect on yellow nutsedge (*Cyperus esculentus*) at 28 days after treatment; however, the mixture of clomazone + pendimethalin at 1145 g ha^{-1} with 4485 g ha^{-1} propanil showed a neutral response [60]. An antagonistic response occurred in yellow nutsedge used as a control when treated with 760 and 1540 g ha^{-1} of clomazone plus pendimethalin mixed with 1120 or 2240 g ha^{-1} of propanil at 28 DAT; however, 1145 g ha^{-1} of clomazone plus pendimethalin mixed with 4485 g ha^{-1} of propanil resulted in a neutral interaction [61]. Unlike yellow nutsedge, a synergistic response occurred when barnyardgrass was treated with all rates of clomazone plus pendimethalin mixed with either rate of propanil evaluated at 56 days after treatment.

An antagonistic effect of metribuzin with halosulfuron and metribuzin with flumioxazin at the different dose and mixture ratios was observed on common lambsquarters (*Chenopodium album*) and redroot pigweed (*Amaranthus retroflexus*) and in potato biomass. On the other hand, the effect of metribuzin with flumioxazin mixtures was antagonistic on potato maximum quantum efficiency (Fv/Fm) while metribuzin with halosulfuron mixtures followed the additive model on Fv/Fm [62]. The mixture of chloridazon and clopyralid followed additive model on *Portulaca oleracea* L., *Solanum nigrum* L., *Amaranthus retroflexus* L., and *Chenopodium album* L. In contrast, desmedipham, phenmedipham, ethofumesate, and clopyralid mixtures showed a synergistic effect on all species except *P. oleracea* at 80 and 90% response levels. The binary mixture of desmedipham+ phenmedipham+ ethofumesate and chloridazon represented additive effect on *S. nigrum* and *A. retroflexus* and followed an antagonism effect on *C. album* and *P. oleracea* [63]. The greenhouse research investigated by [64] showed the mixtures of mesosulfuron+ iodosulfuron + pinoxaden followed synergism effect on wild oat (*Avena fatua*) and *Phalaris minor.* If oxadiargyl + rimsulfuron and metribuzin + rimsulfuron mixed with (25:75)% mixture ratio, a high reduction of common lambsquarters (*Chenopodium album*) and redroot pigweed (*Amaranthus retroflexus*) provided at potato emergence stage in the field [65].

2. Herbicides with adjuvants

Historically, adjuvants are essential components for herbicide-resistant weeds control. To improve herbicides' performance or application objective, adjuvants are used in the spray tank. These adjuvants are commonly added to the spray tank to improve herbicidal activity or application characteristics [66]. According to the [67] "adjuvants are the substances used with a herbicide to improve its performance." In the last definition, "adjuvants are already included in the formulations of some herbicides available for sale. They may be purchased separately and added into a tank mix before use" [68]. Generally, adjuvants have been developed to assist herbicides. They allow mix and handle with herbicide active ingredients better, contact to target weed, increase droplet coverage, and spray retention and droplet drying [66]. Adjuvants diminish or even eliminate spray application problems [69] (e.g., drift reduction) [70], enhance herbicide cuticle penetration and cellular accumulation [71], and decline herbicide amount and total weed control costs. Furthermore, they lead to a significantly greater herbicide efficacy [72] and consequently a lower total herbicide concentration to achieve a given effect [73], as well as promoting the formulation's ability to kill the targeted species without harming other plants [74]. In terms of environmental aspects, they can decrease herbicide leaching through soil profile [75]. However, adjuvant addition does not significantly improve control in some circumstances. Adjuvants can sometimes exhibit adverse effects such as declined herbicide activity (antagonistic effects) [76], enhanced formulation ability to spread or persist in the unwanted environment [77], and increased harmful effects on nontarget plants and aquatic species [78]. Adjuvants are divided into activators, spray modifiers, and utility modifiers [79]. Activators are components that change characteristic herbicides such as viscosity and particle size, evaporation, etc. They improved herbicide activity, spread, absorption into a tissue, rainfastness, and reduced herbicide photodegradation. There are three categories of activators: surfactants, wetting agents, and oils [79].

Surfactants are the most widely used and probably the most essential adjuvants [80]. Surfactants can be classified into nonionic, cationic, anionic, and ampholytic based on their ability to ionize the aqueous solution. Organosilicone and silicone surfactants are two types of nonionic surfactants. Cationic surfactants, which have a positive charge, often are not applied with herbicides, and anionic ones are rarely utilized with herbicides. Ampholytic (amphoteric) have both positive and negative charges, that is, in aqueous solution are capable of forming cations or anions. Wetting agents increase solution spread on the leaves [79]. Oils increase herbicide uptake by increasing the time of retention. They mixed with water via emulsifiers. Oils have uniform droplet size (reduction of drift), decreasing spray evaporation and rainfastness time, and increasing penetration into waxy leaves. They can be classified as: crop oils, dormant oils, crop oil concentrates, vegetable oils, vegetable oil concentrate, modified vegetable oil, and modified vegetable oil concentrate. In addition, spray modifiers are among the most important adjuvants, which influence the delivery and placement of spray solution [81]. They limit or alter the physicochemical characteristics of spray solution, make herbicide spray easier to aim, reduce herbicide drift in the air, and cause the spray to adhere to plants more readily. Spray modifiers include thickening agents (i.e., invert emulsions and polymers), stickers, spreaders, spreader stickers, foaming agents, humectants, and UV absorbents. Utility modifiers are the third group of adjuvants, which help minimize handling and application problems. They do not directly improve efficacy, although they widen the conditions in which an herbicide can be used or maintain the integrity of the spray solution. For instance, utility modifiers diminish foaming, promote solubility, modify pH, or decrease spray drift. Emulsifiers, dispersants, cosolvents, ammonium fertilizers, and stabilizing, coupling, compatibility, buffering, and antifoam agents can be addressed as the types of modifiers.

2.1 Review of research on the positive effects of adjuvants mixture herbicides on weeds

Adjuvants can be especially effective in increasing the biological activity of many herbicides [82]. Previous studies reported that density, viscosity, surface tension, contact angle, droplet size, and droplet evaporation of the spray solution could change with the addition of adjuvants to the spray solution [83]. The activity of tribenuron-methyl significantly enhances following the use of NIS (20% isodecyl alcohol ethoxylate + 0.7% silicone surfactants), an anionic surfactant (25.5% alkyl ether sulfate sodium salt), and vegetable oil (95% natural rapeseed oil with 5% compound emulsifiers) on *Sinapis arvensis*, *Tripleurospermum inodorum*, *Papaver rhoeas*, and *C. album*. Further, only minor differences are observed among the tested adjuvant [84]. The character of foliar surfaces such as cuticle, stomata and trichomes number, leaf position, angle, and leafage is different in various weed species that affect retention and deposition of herbicides [85]. COC (crop oil concentrate), NIS, MSO (methylated soybean oil), and COC-DRA (crop oil concentrate-drift retardant adjuvant) with lactofen increased the spray solution viscosity by 4.3, 2.6, 3.6, 7.5, respectively. Lactofen containing COC, NIS, MSO (methylated soybean oil), and COC-DRA increased viscosity by 4.3%, 2.6%, 3.6%, and 5.7%, respectively, compared with lactofen alone [86]. Methylated seed oil (MSO) and NIS promote the foliar absorption and efficacy of many herbicides such as primisulfuron, rimsulfuron, imazethapyr, quinclorac, and several graminicides for grass weed control [87]. Nonionic surfactants improve glyphosate absorption by 20 times greater, and spray drop is spread 200-fold more than when no adjuvant is added [88]. Furthermore, some researchers reported the strong effect of mineral and vegetable oil on clodinafop-propargyl and diclofop-methyl + fenoxaprop-p-ethyl on *Lolium multiflorum*, *Avena ludoviciana*, and *Phalaris minor* [89]. Seed-oil-based crop oils and organosilicone adjuvants combined with halosulfuron lead to 100% control of *Cyperus rotundus* L. at 8 weeks after treatment (WAT) compared with a combination of halosulfuron with the nonionic or paraffin-based crop oil adjuvants (<90% control) [90]. The measurement of ED_{50} and ED_{90} showed that Citogate (0.1 and 0.2%) increased sulfosulfuron efficacy [91].

Generally, environmental agents affect the efficacy of the mixture of herbicides with adjuvants. In the mixture, rain shortly after utilizing herbicides is among the most detrimental issues for performance. Given that the rainfastness of herbicides increases by applying adjuvants, the effect should be considered when selecting an adjuvant [92]. A study [93] represented a shorter critical rain-free period following the addition of an OSL adjuvant to glyphosate. This decline can be attributed to the lower liquid surface tension of glyphosate caused by the OSL (Organosilicone) adjuvant and the subsequent promotion of the stomatal infiltration of glyphosate into the plant. The conventional adjuvants produced slower absorption of the 14C-glyphosate, as the maximum absorption was not achieved until at least 24 h in redroot pigweed, remaining similar until 72 h [88]. The effect of the vegetable oil on tribenuron-methyl's rainfastness was significantly lower than that of the surfactants with rain at

1 h, while no significant differences among the three adjuvants were observed when rain occurred at 2 and 4 h [84].

2.2 No or negative interaction between herbicides and adjuvants

Adjuvants can significantly enhance the effect of an herbicide, while they fail to increase control and cause harmful effects on nontarget plants in some circumstances (antagonistic effect). Several studies have revealed that *A. theophrasti* is more controlled by adding AMS (ammonium sulfate) into herbicides; however, the control of other species such as *C. album* is not always improved [94]. The combination of sethoxydim and halosulfuron with COC or MSO is antagonistic to smooth crabgrass (*Digitaria ischaemum* (Schreb.) ex Muhl.) [76]. Flumioxazin does not damage wheat or cabbage except after adding silicone adjuvant, which enhances the retention of the spray solution [95]. Adjuvant addition slows down degradation and elevates the level of phenmedipham residue in the soil [77]. The addition of nonionic surfactants to dicamba plus glyphosate tank mixture not only decreased contact angle and surface tension but also droplet size [96].

3. Risk assessment of mixture herbicides on soil: emphasis on earthworm (*Eisenia fetida* L.)

Continuous application of herbicides may lead to soil pollution and affect soil fauna [97]. Generally, herbicides applied alone and in mixture negatively influenced nottargeted animals [98]. As soil inhabitant animals, earthworms might be affected, although the site of action herbicides is not targeted toward animals. They are bioindicators for determining herbicide and heavy metals pollution in soil due to their high sensitivity to soil pollution [99, 100]. The *Eisenia fetida* is currently used as test species in ecotoxicology [101]. There are many methods of testing the toxicity of chemicals to earthworms. Tests include two kinds: a paper contact toxicity and an artificial soil test. A simple paper contact toxicity test is described as an optional initial screen to indicate those substances likely to be toxic to earthworms in soil and which will require further more detailed testing in artificial soil. The artificial soil test gives toxicity data more representative of the natural exposure of earthworms to chemicals [102]. On the base of LC_{50}, for the contact test, the concentration of the test substance is expressed in mg cm^{-2}. For the artificial soil test, it is expressed in mg kg^{-1} (dry weight). The LC_{50} of a reference substance should be occasionally determined to ensure that the laboratory test conditions are adequate and have not changed significantly. Only contact filter paper and artificial soil tests adopt mortality (LC_{50}) as the toxic endpoint in all acute toxicity test methods and have received the most attention. The screening test (filter paper contact test) involves exposing earthworms to test substances on moist filter paper to identify potentially toxic chemicals to earthworms in the soil. The artificial soil test involves keeping earthworms in samples of precisely defined artificial soil to which a range of concentrations of the test substance has been applied. Mortality is assessed 7 and 14 days after application. One concentration resulting in no mortality and one resulting in total mortality should be used. The mortality in the controls should not exceed 10% at the end of either test. Only contact filter paper and artificial soil tests exposure protocols using mortality (LC_{50}) as the toxic endpoint and *E. fetida* as the test species have received the most attention, with the latter being adopted by both [101] and European Economic Community [102] in Europe and the United States Environmental Protection Agency in the United States.

As mentioned before, additive, synergism, and antagonism are three types of herbicide interactions. Concentration addition (CA) and independent action (IA) are two common reference models for determining mixture toxicity.

3.1 Concentration addition (CA)

The toxicity of herbicide mixtures with a similar mode of action is estimated by concentration addition (CA) [103], which has extensively been used for herbicides, and is most straightforward [104]. Generally, CA assumes additivity of toxicity that components will not interact with each other in the mixtures, and the relative potency is equal to the sum of singly potencies [105].

3.2 Independent action

The independent action model (IA) is used for components with the dissimilar mode of action on the organism. They act independently. The toxicity of the total mixture is calculated by the expected effects of each component [106].

3.3 Interaction models

Physical, chemical, and biological interactions of herbicides do not account for by CA and IA models. MIXTOX is an empirical model that determines how much mixture toxicity results deviate from CA and IA model predictions [107]. MIXTOX considered a difference between synergism and antagonism based on concentration and mixture ratios along with deviations [108]. Therefore, experimental design for MIXTOX is considerable due to covering all concentration and mixture ratios [109]; to date, MIXTOX has been used with binary mixture toxicity [110]. The median-effect/combination index (CI) is a method used by [111] to expound chemical interactions. It quantitatively determines the mixtures interactions at various concentrations and mixtures ratios. Pollution interaction is developed by [112].

The response to toxic exposure of *E. fetida* in artificial soil and filter paper tests was estimated using the median-effect equation, as described by [112]:

$$\frac{f_a}{f_u} = \left(\frac{D}{D_m}\right)^m \tag{13}$$

where D is the concentration, Dm is the concentration for 50% effect (50% mortality rate), f_a is the fraction affected by concentration D, f_u is the unaffected fraction ($f_a = 1 - f_u$), and m is the coefficient of the sigmoidicity of the dose-response curve: $m = 1$, $m > 1$, and $m < 1$ indicate hyperbolic, sigmoidal, and negative sigmoidal dose-response curves, respectively. Therefore, the method considers both the potency (D_m) and shape (m) parameters. If Eq. (14) is rearranged, then:

$$\mathrm{D} = D_m (f_a (1 - f_a))]^{1/m} \tag{14}$$

The D_m and m values for each pesticide are easily determined by the median-effect plot: x = log (D) versus y = log (f_a/f_u) which is based on the logarithmic form of

Eq. (14). The median effect plot, m is the slope, and log (D_m) is the x-intercept. The conformity of the data to the median-response principle can be readily manifested by the linear correlation coefficient (r) of the data to the logarithmic form of Eq. (14).

These parameters were then used to calculate concentrations of the pesticides and their combinations required to produce various effect levels according to Eq. (14); combination index (CI) values were then calculated according to the general combination index equation for n chemical combination at 10%, 50%, and 90% mortality rate:

$$(CI)_X = \sum_{j=1}^{n} \frac{(D)_j}{(D_x)_j} = \sum_{j=1}^{n} \frac{(D_x)_{1-n}\{\frac{[D]_j}{\sum_1^n [D]}}{(D_x)_j\ \{\frac{(f_{ax})_j}{1-(f_{ax})_j}} \tag{15}$$

where $^n(CI)_x$ is the combination index for n chemicals at x% effect level; (Dx)1_n is the sum of the concentration of n pesticides causing x% mortality rate of the earthworms in the mixture, $\frac{[D]_j}{\sum_1^n [D]}$ is the proportionality of the concentration of each of n pesticides causing x% mortality rate in combination; and $(D_x)_j\ \{\frac{(f_{ax})_j}{1-(f_{ax})_j}$ is the concentration of each pesticide causing x% mortality rate. From Eq. (15), CI < 1, CI = 1, and CI > 1 indicate synergism, concentration addition, and antagonism, respectively. Where c_{mix} and E (c_{mix}) are the total concentration and total effect of the mixture, respectively. E (ci) denotes the effect of the ith component with the concentration of ci in the mixture.

$$(EC)_{X,mix} = (\sum_{i=1}^{n} \frac{p_i}{EC_{x,i} \times CI_{x\ comp}} \tag{16}$$

CIx comp is the computed combination index value for the mixture at the x level of effect (x%) from the experimental toxicity curve of the mixture [113].

3.4 Review of research on the effect of mixtures of herbicides on *Eisenia fetida*

The study of herbicide mixtures on *Eisenia fetida* is rare. The (50:50) and (25:75)% mixture ratios of metribuzin plus halosulfuron and metribuzin plus flumioxazin provided higher toxicity than the other mixture ratios (100:0) and (0:100)% on earthworm biomass, respectively. Isobologram demonstrated metribuzin plus halosulfuron and metribuzin plus flumioxazin followed an antagonistic effect meaning that the mixtures retracted the action of the herbicide in the earthworms relative to a concentration addition (CA) reference model. Earthworms exposed to a mixture of metribuzin plus halosulfuron and metribuzin plus flumioxazin showed that increased exposure time decreased the LC_{50} in filter paper and artificial soil tests on *Eisenia fetida* mortality. The binary mixture experiments demonstrated for both experiments an apparent antagonistic effect on two types of tests [114]. Antagonistic effects are detected from many mixtures because the compounds in the mixture may stimulate the metabolism of each other, leading to affected absorption in the organism [115]. Synergistic effects become significantly dangerous to soil organisms once the mixture toxicity is much greater than its predicted level [116]. Principles of concentration addition model to assess the impact of triazine herbicides on organophosphate insecticide toxicity to the earthworm *Eisenia fetida*. Atrazine and cyanazine also increased the toxicity of chlorpyrifos 7.9- and 2.2-fold, respectively. However, simazine caused no toxicity to the worms and did not affect chlorpyrifos toxicity in binary mixture experiments. The uptake of chlorpyrifos into the worms was reduced when found in binary mixtures with atrazine, so an increased uptake cannot be considered an explanation. The synergistic effects might be linked to increased biotransformation of the original phosphorus-sulfur bond into a phosphorus-oxygen bond characteristic of oxon derivatives [117]. Atrazine disrupts photosynthesis, which may induce cytochrome P_{450} and general esterase activities in *E. fetida* [117]. Cytochrome P_{450} has an essential role in metabolism [5, 118]. These enzymes break down pesticides by either increasing or decreasing the toxicity of other pesticides depending on whether the resulting metabolites are more or less toxic than their parent compounds [119].

Several herbicides (acetochlor, anilofos, flutamone, pretilachlor, S-metolachlor, and terbutryn) were very toxic in contact toxicity but were low in soil toxicity testing [120]. The mixture of tribenuron methyl (TBM) plus tebuconazole (TEB) showed an antagonistic effect on the earthworms in filter paper and artificial soil tests. In the chronic toxicity experiment, both high concentrations of TBM and TEB, single or combined, induced oxidant stress in the earthworms, and the cellulase activity was inhibited in the earthworm exposed to high concentrations of TBM at the early 35 exposure period. However, both pesticides did not damage the DNA of earthworms in all treatments [99]. Both acute and chronic toxicity tests play an essential role in the risk evaluation of pesticides to earthworms. They are considered valuable for predicting the responses of soil organisms to pesticides [121]. An antagonistic effect was observed the binary mixture of butachlor plus λ-cyhalothrin at all effect levels in artificial soil test, while it shows synergism effect in filter paper test [122]. In the research of Chen et al., [122], the binary mixture of butachlor plus atrazin showed moderate synergism at the highest effect levels. An additive and slightly synergism were observed at $< 0.2fa$ in artificial soil test. The mixtures of atrazine plus exhibited a synergism response in filter paper and artificial soil tests on *Eisenia fetida* mortality. Yang et al. [123] reported the combination of acetochlor plus chlorpyrifos followed a synergism response at 4:1 and 3:2 combination. An antagonistic response was observed the combination of 2:3 and 1:4 of clothianidin plus acetochlor, while a dual additive/antagonist response showed at 4:1, 1:1, and 3:2 combination on *Eisenia fetida* mortality. The most strongly synergistic reported at phoxim plus butachlor plus λ-cyhalothrin combination at the all range. The mixture of atrazin plus butachlor plus cadmium exhibited a slight synergism on *Eisenia fetid* mortality [124].

4. Conclusion

Herbicide resistance is a pervasive challenge in intensive agriculture. Applying multiple modes of action can help to manage herbicide-resistant weeds. Herbicide mixture is a powerful tool to prevent, delay, and control herbicide-resistant weeds. The choice of the most appropriate mixture is crucial and is based on herbicide components, formulation, and weed species. The reference models used to determine the interaction of herbicide and the use of isobologram can illustrate the synergism, additive, and antagonism responses by the ED scale. Another method to manage herbicide-resistant weeds is utilizing adjuvant. Adjuvants are the best tool for improving herbicide performance and optimizing herbicide application. In addition, the adjuvant can overcome antagonist response in the tank mixture. Despite the positive effect, the synergism response in high doses can influence the soil animals such as earthworms. Therefore, growers need knowledge of the management strategies to maximize the long-term benefits of herbicide mixture and reduce weed shifts to difficult-to-control and herbicide-resistant weeds.

Acknowledgements

This paper was supported by the Faculty of Agriculture and Natural Resources, University of Mohaghegh Ardabili, Iran, and Mississippi State University, USA, for financial support. This work is supported by the National Institute of Food and Agriculture, U.S. Department of Agriculture, Hatch project under accession number 230100.

Conflicts of interest

The authors declare no conflict of interest.

Author details

Mohammad Taghi Alebrahim[1]*, Elham Samadi Kalkhoran[1] and Te-Ming Paul Tseng[2]

1 Faculty of Agriculture and Natural Resources, Department of Plant production and Genetics, University of Mohaghegh Ardabili, Ardabil, Iran

2 Department of Plant and Soil Sciences, Mississippi State University, Mississippi State, MS, USA

*Address all correspondence to: m_ebrahim@uma.ac.ir

References

[1] Maria-Morales MM, Ventura-Camargo BDC, Hoshina MM. Toxicity of herbicides: Impact aquatic and soil biota and human health. In: Herbicides-Current Reasearch and Case Studies in Use. Janeza, Trdine and Rijeka, Croatina, UK: IntechOpen; 2013. Chapter 16. DOI: 10.5772/55851

[2] Mehdizadeh M, Alebrahim MT, Roushani M, Streibig JC. Evaluation of our different crops' sensitivity to sulfosulfuron and tribenuron methyl soil residues. Acta Agriculturae ScandinavicaSection B—Soil & Plant Science. 2016:706-713. DOI: 10.1080/09064710.2016.1212919

[3] Mehdizadeh M, Alebrahim MT, Roushani M. Determination of two sulfonylurea herbicides residues in soil environment using HPLC and phytotoxicity of these herbicides by lentil bioassay. Bulletin Environmental Contamination Toxicology. 2017; **99**:93-99. DOI: 10.1007/s00128-017-2076-8

[4] Abbas T, Nadeem MA, Tanveer A, Ahmad R. Identifying optimum herbicide mixtures to manage and avoid fenoxaprop-p-ethyl resistant phalaris minor in wheat. Planta Daninha. 2016; **34**(4):787-793. DOI: 10.1590/S0100-83582016340400019

[5] Alebrahim MT, Zangouenejad R, TMP T. Biochemical and molecular knowledge about developing herbicide-resistant weeds. In: Herbicide Resistance in Weeds and Crops. Janeza, Trdine and Rijeka, Croatina, UK: IntechOpen; 2017. Chapter 5

[6] Sabet Zangeneh H, Mohammaddust Chamanabad HR, Zand E, Asghari A, Alamisaeide K, Travlos IS, et al. Cross- and multiple herbicide resistant *Lolium rigidum* Guad. (rigid ryegrass) biotypes in Iran. Journal Agricultural Science Technology. 2018;**20**:1187-1200

[7] Zangouenejad R, Alebrahim MT. Shredded date palm (*Phoenix dactylifera* L.) leaves and cereal straws as much material vs. herbicide options for weed suppression in processing tomato. International Journal of Pest Management. 2021. DOI: 10.1080/09670874.2021.1943050

[8] Northworthy JK, Ward SM, Shaw DR, Llewellyn RS, Nichols RL, Webster TM, et al. Reducing the risks of herbicide resistance: best management practices and recommendations. Weed Science. 2012;**60**(SI 1):31-62. DOI: 10.1614/WS-D-11-11-00155.1

[9] Hanifezade S, Alebrahim MT, Fakhari R. The Effect of rimsulfuron and metribuzin mixture on weed control and antioxidant enzyme of potato. Weed Research Journal. 2017; **9**(2):31-41. (In Persian with English summary)

[10] Beckie HJ, Reboud X. Selecting for weed resistance: Herbicide rotation and mixture. Weed Technology. 2009;**23**: 363-370. DOI: 10.1614/WT-09-008.1

[11] Busi R, Powles SB, Beckie HJ, Renton M. Rotations and mixtures of soil-applied herbicides delay resistance. Pest Management Science. 2020;**76**: 487-496. DOI: 10.1002/ps.5534

[12] Khalil Tahmasebi B, Alebrahim MT, Roldán-Gómez R, Martins da Silveira H, Bianco de Carvalho L, Alcántara-de la Cruz R, et al. Effectiveness of alternative herbicides on three Conyza species from Europe with and without glyphosate resistance. Crop Protection. 2018;

2018(112):350-355. DOI: 10.1016/j.cropro.2018.06.021

[13] Heap I. The international survey of herbicide-resistant weeds. Available from: www.weedscience.org [Accessed: 20 July 2020]

[14] Webster TM. Southern Weed Science Society Weed Survey. In: Proceedings of the Southern Weed Science Society Annual Meeting. Charleston, SC: Southern Weed Science Society; 2012. pp. 267-288

[15] Fish JC, Webster EP, Blouin DC, Bond JA. Imazethapyr co-application interactions in imidazolinone-resistant rice. Weed Technology. 2015;**29**:689-696. DOI: 10.1614/WT-D-15-00030.1

[16] Wrubel RP, Gressel J. Are herbicide mixtures useful for delaying the rapid evolution of resistance? A case study. Weed Technology. 1994;**8**:635-648. DOI: 10.1017/S0890037X00039828

[17] Carlson TP, Webster EP, Salassi ME, Hensley JB, Blouin DC. Imazethapyr plus propanil programs in imidazolinone-resistant rice. Weed Technology. 2011; **25**:204-211. DOI: 10.1614/WT-D-10-00118.1

[18] Blouin D, Webster E, Bond J. On a method of analysis for synergistic and antagonistic joint-action effects with fenoxaprop mixtures in rice (*Oryza sativa*). Weed Technology. 2010;**24**: 583-589. DOI: 10.1614/WT-D-10-00025.1

[19] Rustom SY, Webster EP, Blouin DC, McKnight BM. Interactions between quizalofop-p-ethyl and acetolactate synthase-inhibiting herbicides in acetyl-coA carboxylase inhibitor-resistant rice production. Weed Technology. 2018;**32**: 297-303. DOI: 10.1017/wet.2018.15

[20] Rustom SY, Webster EP, McKnight BM, Blouin DC. Interactions of quizalofop- p-ethyl mixed with contact herbicides in ACCase-resistant rice production. Weed Technology. 2019;**33**:233-238. DOI: 10.1017/wet.2018.116

[21] Ikeda FS. Resistência de plantas daninhas em soja resistente ao glifosato. 515 Informe Agropecuário. 2013; **34**(276):1-8

[22] Damalas CA. Herbicide tank mixtures: common interactions. International Journal Agriculture Biology. 2004;**6**:209-212. DOI: 1560-8530/2004/06-1-209-212

[23] Takano HK, Beffa R, Preston C, Westra P, Dayan FE. Glufosinate enhances the activity of protoporphyrinogen oxidase inhibitors. Weed Science. 2020;**68**:324-332. DOI: 10.1017/wsc.2020.39

[24] Hatzios KK, Penner D. Interactions of herbicides with other agrochemicals in higher plants. Reviews of Weed Science. 1985;**1**:1-63

[25] Zhang J, Allan SH, Susan EW. Antagonism and synergism between herbicides: Trends from previous studies. Weed Technology. 1995;**9**:86-90. DOI: 10.1017/S0890037X00023009

[26] Puckowski A, Stolte S, Wagil M, Markiewicz M, Łukaszewicz P, Stepnowski P, et al. Mixture toxicity of flubendazole and fenbendazole to *Daphnia magna*. Intenational Journal Hygen Environmental Health. 2017; **220**(3):575-582. DOI: 10.1016/j.ijheh.2017.01.011

[27] Mwense M, Wang XZ, Buontempo FV, Horan N, Young A, Osborn D. Prediction of noninteractive mixture toxicity of organic compounds

based on a fuzzy set method. Journal Chemical Information Computer Scientists. 2004;**44**(5):1763-1773. DOI: 10.1021/ci049 9368

[28] Streibig JC, Jensen JE. Actions of herbicides in mixtures. In: Cobb AH, Kirkwood RC, editors. Herbicides and Their Mechanisms of Action. Sheffield, UK: Sheffield Academic Press; 2000. pp. 153-180

[29] Howard GJ, Webster TF. Generalized concentration addition: A method for examining mixtures containing partial agonists. Journal of Theoretical Biology. 2009;**259**: 469-477. DOI: 10.1016/j.jtbi.2009.03.030

[30] Chou TC. Drug combination studies and their synergy quantification using the Chou–Talalay method. Cancer Research. 2010;**70**:440-446. DOI: 10.1158/0008-5472.CAN-09-1947

[31] Wu J, Tracey L, Davidoff AM. Assessing interactions for fixed-dose drug combinations in tumor xenograft studies. Journal of Biopharmaceutical Statistics. 2012;**22**:35-543. DOI: 10.1080/10543406.2011.556285

[32] Finney DA. Probit Analysis. Cambridge: Cambridge University Press; 1971. DOI: 10.1002/jps.2600600940

[33] Ritz C, Streibig JC, Kniss A. How to use statistics to claim antagonism and synergism from binary mixture experiments. Pest Management Science. 2021;**77**(9):3890-3899. DOI: 10.1002/ps.6348

[34] Schmidt JB, Ritz C. Modelling synergistic effects of appetite-regulating hormones. Synergy. 2016;**3**:1-2. DOI: 10.1016/j.synres.2015.12.001

[35] Colby SR. Calculating synergistic and antagonistic responses of herbicide combinations. Weeds. 1967;**15**:20-22

[36] Reed JD, Keeling JW, Dotray PA. Palmer amaranth (*Amaranthus palmeri*) management in GlyTol® LibertyLink® cotton. Weed Technology. 2014;**28**: 592-600. DOI: 10.1614/WT-D-14-00029.1

[37] Abdelbasit K, Plackett R. Experimental design for joint action. Biometrics. 1982;**38**:171-179. DOI: 10.2307/2530300

[38] Ritz C, Jensen SJ, Gerhard D, Streibig JC. Dose-Response Analysis Using R. New York, NY: Chapman and Hall/CRC; 2019. DOI: 10.1201/b21966

[39] Tallarida RJ. Drug Synergism and Dose-Effect Data Analysis. Chapman & Hall/CRC: Boca Raton, FL; 2000. DOI: 10.1201/9781420036107

[40] Seeger B, Klawonn F, Nguema BB, Steinberg P. Mixture effects of estrogenic pesticides at the human estrogen receptor α and β. Plos One. 2016;**11**:e0147490. DOI: 10.1371/journal.pone.0147490

[41] Holland-Letz T, Leibner A, Kopp-Schneider A. Modeling dose-response functions for combination treatments with log-logistic or Weibull functions. Archives of Toxicology. 2020;**94**: 197-204. DOI: 10.1007/s00204-019-02631-2

[42] Grabovsky Y, Tallarida RJ. Isobolographic analysis for combinations of a full and partial agonist: Curved isoboles. Journal Pharmacology and Experimental Therapeutics. 2004; **310**:981-986. DOI: 10.1124/jpet.104.067264

[43] Ezechiáš M, Cajthaml T. New insight into isobolographic analysis for combinations of a full and partial agonist: Curved isoboles. Toxicology. 2018;**402–403**:9-16. DOI: 10.1016/j.tox.2018. 04.004

[44] Brinkmann M, Hecker M, Giesy JP, Jones PD, Ratte HT, Hollert H, et al. Generalized concentration addition accurately predicts estrogenic potentials of mixtures and environmental samples containing partialagonists. Toxicology In Vitro. 2018;**46**:294-303. DOI: 10.1016/j.tiv.2017.10.022

[45] Cedergreen N, Streibig JC. Can the choice of endpoint lead to contradictory results of mixture toxicity experiments? Environmental Toxicology Chemistry. 2005;**24**:1676-1683. DOI: 10.1897/04-362R.1

[46] Moyson S, Town RM, Vissenberg K, Blust R. The effect of metal mixture composition on toxicity to C. elegans at individual and population levels. Plos One. 2019;**14**:e0218929. DOI: 10.1371/journal.pone.0218929

[47] Sørensen H, Cedergreen N, Streibig JC. A random-effects model for binary mixture toxicity experiments. Journal of Agricultural, Biological Environmental Statistics. 2010;**15**: 562-577. DOI: 10.1007/s13253-010-0041-7

[48] Damalas CA, Eleftherohorinos IG. Dicamba and atrazine antagonism on sulfonylurea herbicides used for johnsongrass (*Sorghum halepense*) control in corn (*Zea mays*). Weed Technology. 2001;**15**:62-67. DOI: 10.1614/0890-037X(2001)015[0062:DAAAOS]2.0.CO;2

[49] Rhodes GN, Jr. Coble HD. Influence of application variables on antagonism between sethoxydim and bentazon. Weed Science. 1984;**32**: 436-441. DOI: 10.1017/S0043174500059294

[50] Zhang W, Webster EP, Blouin DC, Leon CT. Fenoxaprop interactions for barnyardgrass (*Echinochloa crus-galli*) control in rice. Weed Technology. 2005; **19**:293-297. DOI: 10.1614/WT-03-250R1

[51] Bethke RK, Molin WT, Sprague C, Penner D. Evaluation of the interaction between glyphosate and glufosinate. Weed Science. 2013;**61**:41-47. DOI: 10.1614/WS-D-12-00031.1

[52] Gardner AP, York AC, Jordan DL, Monks DW. Glufosinate antagonizes postemergence graminicides applied to annual grasses and johnsongrass. Journal Cotton Science. 2006;**10**:319-327. DOI: 10.1614/WS-D-12-00031.1

[53] Burke IC, Askew SD, Corbett JL, Wilcut JW. Glufosinate antagonizes clethodim control of goosegrass (*Eleusine indica*). Weed Technology. 2005;**19**:664-668. DOI: 10.1614/WT-04-214R1.1

[54] Eytcheson AN, Reynolds DB. Barnyardgrass (*Echinochloa crus-galli*) control as affected by application timing of glufosinate applied alone or mixed with graminicides. Weed Technology. 2019;**33**:272-279. DOI: 10.1017/wet.2018.89

[55] Green JM. Herbicide antagonism at the whole plant level. Weed Technology. 1989;**3**:217-226. DOI: 10.1017/S0890037X00031717

[56] Shimabukuro RH, Walsh WC, Hoerauf RA. Reciprocal antagonism between the herbicides diclofop–methyl and 2,4–D in corn and soybean tissue culture. Plant Physiology. 1986;**80**: 612-617. DOI: 10.1104/pp.80.3.612

[57] Meyer CJ, Norsworthy JK, Kruger GR. Antagonism in mixtures of glufosinate glyphosate and glufosinate clethodim on grasses. Weed Technology. 2021;**35**(1):12-21. DOI: 10.1017/wet.2020.49

[58] Liebl R, Worsham AD. Effect of chlorsulfuron on diclofop phytotoxicity to italian ryegrass (*Lolium multiflorum*). Weed Science. 1987;**35**:383-387. DOI: 10.1017/S0043174500053868

[59] Duus J, Kruse ND, Streibig JC. Effect of mesotrione and nicosulfuron mixtures with or without adjuvants. Plant Daninha. 2018;**36**. DOI: 10.1590/S0100-83582018360100116

[60] Cobb AH. Herbicides and Plant Physiology. London: Chapman & Hal; 1992

[61] Osterholt MJ, Webster EP, McKnight BM, Blouin DC. Interactions of clomazone plus pendimethalin mixed with propanil in rice. Weed Technology. 2021;**35**(5):675-680. DOI: 10.1017/wet.2021.3

[62] Samadi Kalkhoran E, Alebrahim MT, Mohammaddust Chamn Abad HR, Streibig JC, Ghavidel A, Tseng TMP. The joint action of some broadleaf herbicides on potato (*Solanum tuberosum* L.) weeds and photosynthetic performance of potato. Agriculture. 2021;**11**(11):1103. DOI: 10.3390/agriculture11111103

[63] Chitband AA, Ghorbani R, Rashed Mohassel MH, Nabizade M. Joint action of some broadleaf herbicides in sugar beet. International Journal of Pest Management. 2019;**67**(3): 179-186. DOI: 10.1080/09670874.2019.1609128

[64] Kargar M, Ghorbani R, Rashed Mohassel MH, Rastgoo M. Isobolographic analysis for mixture effects of esosulfuron-methyl + iodosulfuron with pinoxaden in wheat (*Triticum aestivum*). Plant Protection. 2019;**4**(30):610-621. DOI: 10.22067/jpp.v30i4.50213

[65] Hanifezade Erdi S, Alebrahim MT, Majd R, Samadi KE. Effect of application ratio and rimsulfuron, oxadiargyl and metribuzin combination application time on weed biomass and tuber yield of potato (*Solanum tuberosum*). Iranian Journal of Weed Science. 2019;**15**(2):79-92. DOI: 10.22092/IJWS.2019.1502.07 (In persian with Englisg abstract)

[66] Curran WS, McGlamery MD, Liebl RA, Lingenfelter DD. Adjuvants for Enhancing Herbicide Performance. Agronomy Facts 37. Pennsylvania, PA. Available from: http://cropsoil.psu.edu/extension/facts/uc106.pdf 2015: The Pennsylvania State University University Park; [Accessed: 22 January 2015]

[67] Ferrell JA, Macdonald GE, Sellers B. Adjuvants. Agronomy Department, Florida Cooperative Extension Service, Institute of Food and Agricultural Sciences, University of Florida. Available from: http://edis.ifas.ufl.edu; 2015 [Accessed: 17 January 2015]

[68] Pringnitz B. Clearing up Confusion on Adjuvants and Additives. Iowa State University Extension Agronomy. Available from: http://www.weeds/iastate.edu/mgmt/qtr98-2/cropoils.htm; 2015 [Accessed: 20 January 2015]

[69] Hock WK. Horticultural Spray Adjuvants. College of Agricultural Sciences. Cooperative Extension Pennsylvania Department of Agriculture Pennsylvania State University. Available from: http://pubs.cas.psu.edu/FreePubs/

pdfs/uo202.pdf; 2015 [Accessed: 21 January 2015]

[70] Celen IH. The effect of spray mix adjuvants on spray drift. Bulgarian Journal of Agricultural Science. 2010; **16**(1):105-110

[71] Bruce JA, Carey JB, Penner D, Kells JJ. Effects of growth stage and environment on foliar absorption, translocation, and activity of Nicosulfuron in Quackgrass (*Elytrigia repens*). Weed Science. 1996;**44**(3): 447-454

[72] Bunting JA, Sprague CL, Riechers DE. Proper adjuvant selection for foramsulam activity. Crop Protection. 2004;**23**(4):361-366. DOI: 10.1016/j.cropro.2003.08.022

[73] Underwood AL. Adjuvant trends for the new millennium. Weed Technology. 2000;**14**(4):765-772. DOI: 10.1614/0890-037X(2000)014[0765:ATFTNM]2.0.CO;2

[74] Hess FD, Foy CL. Interaction of surfactants with plant cuticles. Weed Technology. 2000;**14**(4): 807-813. DOI: 10.1614/0890-037X(2000)014[0807:IOSWPC]2.0.CO;2

[75] Reddy KN. Effect of acrylic polymer adjuvants on leaching of bromacil, diuron, norflurazon and simazine in soil columns. Bulletin Environment Contamination Toxicology. 1993;**50**(3):449-457. DOI: 10.1007/BF00197207

[76] Kammler KJ, Walters SA, Young BG. Effects of adjuvants, halosulfuron, and grass herbicides on *Cucurbita* spp. injury and grass control. Weed Technology. 2010;**24**(2):147-152. DOI: 10.1614/WT-D-09-00015.1

[77] Kucharski M. Degradation of phenmedipham in soil under laboratory conditions. Vegetable Crops Research Bulletin. 2004;**60**(2):63-70

[78] Tyler MJ. Environmentally friendly: A false sense of security species. Newsletter of the Species Survival Commission, IUCN. The World Conservation Union. 1997;**29**:20-21

[79] Pacanoski Z. Herbicides and adjuvants. In: Herbicides Physiology Action and Safety. 2015Injury and Grass Control. DOI: 10.5772/60842.

[80] Miller PA, Westra P. How surfactants work, No. 0.564. Colorado State University Cooperative Extension, Crop Fact Sheet; 1998. Available from: http://thanyagroup.com/research/download/25530111_2.pdf [Accessed: 17 January 2015]

[81] Hall FR, Chapple AC, Downer RA, Kirchner LMJ, Thacker RM. Pesticide application as affected by spray modifiers. Pesticide Science. 1993;**38**(2–3):123-133. DOI: 10.1002/ps.2780380207

[82] Green JM, Beestman GB. Recently patented and commercialized formulation and adjuvant technology. Crop Protection. 2007;**26**(3):320-327. DOI: 10.1016/j.cropro.2005.04.018

[83] Cunha J, Alves G. Características físico-químicas de soluções aquosas com adjuvantes de uso agrícola. Interciencia. 2009;**34**:655-659

[84] Pannacci E, Mathiassen KS, Kudsk P. Effect of adjuvants on the rainfastness and performance of tribenuron-methyl on broad-leaved weeds. Weed Biology and Management. 2010;**10**(2):126-131. DOI: 10.1111/j.1445-6664.2010.00376.x

[85] Kraemer T, Hunsche M, Noga G. Surfactant-induced deposit structures in

relation to the biological efficacy of glyphosate on easy- and difficult-to-wet weed species. Pest Management Science. 2009;**65**:844-850. DOI: 10.1002/ps.1759

[86] Moraes JG, Luck JD, Antuniassi UR, Hoffmann WC, Kruger GR. Effect of adjuvants on physical properties of glyphosate and PPO-inhibiting herbicide spray mixtures. In: Pesticide Formulation and Delivery Systems: 39th Volume, Innovative Formulation, Application and Adjuvant Technologies for Agriculture. West Conshohocken, PA, USA: ASTM International; 2019. pp. 64-74. DOI: 10.1520/STP161920180130

[87] Hutchinson PJS, Eberlein CV, Tonks DJ. Broadleaf weed control and potato crop safety with postemergence rimsulfuron, metribuzin, and adjuvant combinations. Weed Technology. 2004;**18**(3):750-756. DOI: 10.1614/WT-03-172R1

[88] Singh D, Singh M. Absorption and translocation of glyphosate with conventional and organosilicone adjuvants. Weed Biology and Management. 2008;**8**(2):104-111. DOI: 10.1111/j.1445-6664.2008.00282.x

[89] Stagnari F, Onofri A, Covarelli G. Influence of vegetable and mineral oils on the efficacy of some post-emergence herbicides for grass weed control in wheat. Journal of Pesticide Science. 2006;**31**(3):339-343

[90] McDaniel GL, Klingeman WE, Witte WT, Flanagan PC. Choice of adjuvant with halosulfuron affects purple nutsedge control and nursery crop tolerance. HortScience. 2001;**36**(6):1085-1088

[91] Mehdizadeh M, Alebrahim MT. Effect of some adjuvants application on enhancing sulfosulfuron herbicide performance on *Phalaris minor*—Poaceae. Azarian Journal of Agriculture. 2015;**2**(1):7-11

[92] Kudsk P, Mathiassen SK. Adjuvant effects on the rainfastness of iodosulfuron-methyl + mesosulfuron formulations. In: Proceedings of the 7th International Symposium on Adjuvants for Agrochemicals. Stellenbosch: Michael North; 2004. pp. 159-164

[93] Reddy KN, Singh M. Organosilicone adjuvant effects on glyphosate efficacy and rainfastness. Weed Technology. 1992;**6**(2):361-365. DOI: 10.1017/S0890037X00034874

[94] Young BG, Knepp AW, Wax LM, Hart SE. Glyphosate translocation in common lambsquarters (*Chenopodium album*) and velvetleaf (*Abutilon theophrasti*) in response to ammonium sulfate. Weed Science. 2003;**51**(2):151-156. DOI: 10.1614/0043-1745(2003)051[0151:GTICLC]2.0.CO;2

[95] Nelson EA, Penner D. Reduction of isoxaflutole injury to corn (*Zea mays* L.) with herbicide safeners and water-repellent adjuvants. Weed Technology. 2006;**20**(4):999-1003. DOI: 10.1614/WT-05-100.1

[96] Ferreira PHU, Thiesen LV, Pelegrini G, Ramos MFT, Pinto MMD, Ferreira MDC. Physicochemical properties, droplet size, and volatility of dicamba with herbicides and adjuvants on tank mixture. Scientific Reports. 2020;**10**:18833. DOI: 10.1038/s41598-021-84555-5

[97] Zhang Q, Saleem M, Wang C. Probiotic strain *Stenotrophomonas acidaminiphila* BJ1 degrades and reduces chlorothalonil toxicity to soil enzymes, microbial communities and plant roots. AMB Express. 2017;**7**:227. DOI: 10.1186/s13568-017-0530-y

[98] Wang JH, Zhu LS, Liu W, Wang J, Xie H. Biochemical responses of earthworm (*Eisenia fetida*) to the pesticides chlorpyrifos and fenvalerate. Toxicology Mechanism and Methods. 2012;**22**(3):236-241. DOI: 10.3109/15376516.2011.640718

[99] Chen J, Saleem M, Wang C, Liang W, Zhang Q. Individual and combined effects of herbicide tribenuron-methyl and fungicide tebuconazole on soil earthworm *Eisenia fetida*. Scientific Reports. 2018;**8**:2967. DOI: 10.1038/s41598-018-21288-y

[100] Hirano T, Tamae K. Earthworms and soil pollutants. Sensors (Basel). 2011; **11**(12):11157-11167. DOI: 10.3390/s111211157

[101] OECD. Guidelines for the Testing of Chemicals No. 222. Earthworm Reproduction Test (*Eisenia fetida/ Eisenia andrei*). Paris: Organization for Economic Co-operation and Development; 2004. pp. 1-9

[102] European Economic Community. 1985

[103] Zhang A, Xuemei X, Ye J, et al. Stereoselective toxicity of malathion and its metabolites, malaoxon and isomalathion. Environmental Chemistry Letters. 2010;**9**(3):369-373. DOI: 10.1007/s10311-010-0288-9

[104] Vaj C, Barmaz S, SØrensen PB, et al. Assessing, mapping and validating site-specific ecotoxicological risk for pesticide mixtures: A case study for small-scale hot spots in aquatic and terrestrial environments. Ecotoxicology Environmental Safety. 2011;**74**: 2156-2166

[105] Deneer JW. Toxicity of mixtures of pesticides in aquatic systems. Pest Management Science. 2000;**56**:516-551

[106] Faust M, Altenburger R, Backhaus T, et al. Predictive assessment of the aquatic toxicity of multiple chemical mixtures. Journal Environmental Quality. 2000;**29**: 1063-1068. DOI: 10.2134/jeq2000.00472425002900040005x

[107] Jonker MJ, Svendsen C, Bedaux JJM, et al. Significance testing of synergistic/antagonistic, dose level-dependent, or dose ratio-dependent effects in mixture doseresponseanalysis. Environmental Toxicology Chemistry. 2005;**24**:2701-2713

[108] Loureiro S, Svendsen C, Ferreira ALG, et al. Toxicity of three binary mixtures to *Daphnia magna*: Comparing chemical modes of action and deviations from conceptual models. Environmental Toxicology Chemistry. 2010;**29**:1716-1726. DOI: 10.1002/etc.198

[109] Rudzok S, Schlink U, Herbarth O, et al. Measuring and modeling of binary mixture effects of pharmaceuticals and nickel on cell viability/cytotoxicity in the human hepatoma derived cell line $HepG_2$. Toxicology Applied Pharmacology. 2010;**244**:336-343. DOI: 10.1016/j.taap.2010.01.012

[110] Rodea-Palomares I, Gonzalez-Pleiter M, Martin-Betancor K, Rosal R, Fernandez-Pinas F. Additivity and interactions in ecotoxicity of pollutant mixtures: Some patterns, conclusions, and open questions. Toxics. 2015;**3**(4): 342-369. DOI: 10.3390/ toxic s3040 342

[111] Gonzalez-Pleiter M, Gonzalo S, Rodea-Palomares I, Leganes F, Rosal R, Boltes K, et al. Toxicity of five antibiotics and their mixtures towards photosynthetic aquatic organisms: Implications for environmental risk assessment. Water Research. 2013;**47**(6): 2050-2064. DOI: 10.1016/j.watres.2013.01.020

[112] Chou TC, Talalay P. Quantitative analysis of dose-effect relationships: The combined effects of multiple drugs of enzyme inhibitors. Advances in Enzyme Regulation. 1984;**22**:27-55. DOI: 10.1016/0065-2571(84)90007-4

[113] Chou TC. Theoretical basis, experimental design, and computerized simulation of synergism and antagonism in drug combination studies. Pharmacology Review. 2006;**58**:621-681. DOI: 10.1124/pr.58.3.10

[114] Samadi Kalkhoran E, Alebrahim MT, Mohammaddust Chamn Abad HR, Streibig JC, Ghavidel A. Investigation of relative toxicity of some combined herbicides on earthworm (*Eisenia fetida* L.) biomass. Iranian Journal of Soil and Water Research. 2021;**52**(6):1661-1672. DOI: 10.1124/pr.58.3.10

[115] Hernández AF, Parrón T, Tsatsakis AM, Requena M, Alarcón R, López-Guarnido O. Toxic effects of pesticide mixtures at a molecular level: Their relevance to human health. Toxicology. 2013;**307**:136-145. DOI: 10.1016/j.tox.2012.06.009

[116] Denton DL, Wheelock CE, Murray SA, Deanovic LA, Hammock BD, Hinton DE. Joint acute toxicity of esfenvalerate and diazinon to larval fathead minnows (*Pimephales promelas*). Environmental Toxicology Chemistry. 2003;**22**:336-341. DOI: 10.1002/etc.5620220214

[117] Lydy MJ, Linck SL. Assessing the impact of triazine herbicides on organophosphate insecticide toxicity to the earthworm *Eisenia fetida*. Archives Environment. G. Yang et al. Ecotoxicology and Environmental Safety Contamination Toxicology. 2003;**45**: 343-349. DOI: 10.1007/s00244-002-0218-y

[118] Abdollahi F, Alebrahim MT, Ngov CH, Lallemand E, Lallemand E, Zheng Y, et al. Innate promiscuity of the CYP706 family of P450 enzymes provides a suitable context for the evolution of dinitroaniline resistance in weed. New Phytologist. 2020;**229**(6): 3253-3268. DOI: 10.1111/nph.17126 . hal-03102331

[119] Anderson TD, Zhu KY. Synergistic and antagonistic effects of atrazine on the toxicity of organophosphorodithioate and organophosphorothioate insecticides to Chironomus tentans (Diptera: Chironomidae). Pesticide Biochemistry Physiology. 2004;**80**:54-64. DOI: 10.1016/j.pestbp.2004.06.003

[120] Xiao NW, Jing BB, Ge F, Liu XH. The fate of herbicide acetochlor and its toxicity to *Eisenia fetida* under laboratory conditions. Chemosphere. 2006;**62**: 1366-1373. DOI: 10.1016/j.chemosphere.2005.07.043

[121] Zhang Q, Zhang B, Wang C. Ecotoxicological effects on the earthworm *Eisenia fetida* following exposure to soil contaminated with imidacloprid. Environmental Science Pollutution. 2014;**21**:12345-12353. DOI: 10.1007/s11356-014-3178-z

[122] Chen CH, Wang Y, Zhao X, Qian Y, Wang Q. Combined toxicity of butachlor, atrazine and λ-cyhalothin on the earthworm Eisenia fetida by combination index (CI)-isobologram method. Chemosphere. 2014;**112**: 393-401. DOI: 10.1016/j.chemosphere.2014.04.070

[123] Yang G, Chen CH, Wang Y, Peng Q, Zhao H, Guo D, et al. Mixture toxicity of four commonly used

pesticides at different effect levels to the epigetic earthworm, *Eisenia fetida*. Ecotoxicology and Environmental Safety. 2017;**142**:29-39. DOI: 10.1016/j.ecoenv.2017.03.037

[124] Wang Y, Chen CH, Qian Y, Zhao X, Wang Q. Ternary toxicological interactions of insecticides, herbicides, and a heavy methal on the earthworm *Eisenia fetida*. Journal of Hazardous Materials. 2015;**284**:233-240. DOI: 10.1016/j.jhazmat.2014.11.017

Chapter 5

Integrated Weed Management in Coffee for Sustainable Agriculture – A Practical Brazilian Approach

Daniel Resende Fontes, Andrew de Paula Ribeiro, Marcelo Rodrigues dos Reis, Miriam Hiroko Inoue and Kassio Ferreira Mendes

Abstract

Brazil is the largest coffee exporter in the world market and ranks second among coffee-consuming countries. The use of technology has been largely responsible for the great development of Brazilian agriculture in recent years. Then, the efficiency of integrated weed management has made the country more competitive in coffee farming. Therefore, integrated weed management (IWM) practices are the foundation for sustainable weed management in coffee fields. Weed competition cause losses in crop production. In weed control, besides chemical control, there are other methods that are efficient, economical, and beneficial to the coffee plant and to the environment that can be used on any property, such as preventive and cultural managements; and mechanical, biological, and physical controls. The combination of weed control methods has proven to be a sustainable practice in coffee production. In integrated management, the inherent advantages of each control method must be combined. Lastly, IWM provides an efficient control action with lower costs, better environmental conservation, and higher crop productivity. Thus, this chapter discusses the main practices of sustainable weed management in coffee, addressing issues such as competition, benefits, main weeds, and IWM systems.

Keywords: integrated weed management, weed control, herbicide, cover crop

1. Introduction

Brazil is the largest coffee exporter in the world market and ranks second among coffee-consuming countries. This quantity of coffee corresponds to one-third of the world's production, which places it as the largest producer for more than 150 years. The country has approximately 264,000 coffee-producing farms, of which 78% are considered family coffee farming [1]. Brazilian coffee-producing farms are present in 5 geographic regions, in 16 states of the Federation, in which there are 1448 cities that produce coffee, which corresponds to approximately 26% of Brazilian cities [1]. The Brazilian coffee planted area in 2020 corresponded to 2.162 million hectares, an area that includes the *Coffea arabica* and *Coffea canephora* [2]. Of this total, 276,000 hectares (13%) are in training and 1.885 million hectares (87%) in production [2]. In the case of Brazil, besides the development of technology, the availability of land and labor makes the country internationally competitive. As a technology-intensive crop, coffee is an activity that generates employment and income, especially when considering the other activities throughout the product chain, as well as the trade balance surplus, a factor that favors economic development. Although the area occupied by coffee plantations is not significant in relation to the area explored with other agricultural activities, coffee contributes significantly to Brazilian agribusiness, both economically and socially. Furthermore, it is possible to verify that the area occupied by Brazilian coffee farming had a reduction of approximately 17% in the last 2 decades [2].

Even so, in the last 20 years (2001–2020), the volume of coffee produced increased by approximately 200% as a result of the increase in crop productivity [2]. The use of technology has been largely responsible for the great development of Brazilian agriculture in recent years. In coffee growing, it is no different! Then, the efficiency of the integrated management of pests, diseases, and weeds; the nutrition of coffee trees; pruning and conduction of crops; irrigation, and the development of new varieties have made the country more competitive in coffee farming.

Coffee plants have a very low initial growth rate [3], which also impairs soil cover [4, 5]. Thus, especially during the juvenile phase (up to 2 years in the field), the coffee crop is highly sensitive to competition from weed species [5, 6]. This results in a noticeable reduction in coffee growth and yield, and weed control is one of the major field management practices, which can entail high costs [4, 7, 8]. In Brazil, there are different coffee-producing regions, each using specific cultural practices for crop management [9]. Therefore, the integrated weed management (IWM) practices adopted will vary between farms, depending on local characteristics. In fact, the adoption of site-specific IWM practices is the foundation for sustainable weed management in any cropping system [10]. However, this is not always a usual practice of the grower, often opting for chemical control only using glyphosate-based products.

The objectives of this chapter on IWM of coffee in Brazil are: (a) state the main practices of sustainable weed management and (b) address the major issues of weed competition, benefits, main weed species involved, and discuss the leading IWM systems.

2. Weed competition

Several studies have related the losses in coffee growth when in competition with weeds. In this sense, Oliveira et al. [11] found that without adequate control of weeds, observing the critical periods of control in coffee, there were losses in crop production where the weeds were not controlled throughout the year, reaching reductions of 43%.

It is well known that weeds affect the coffee crop in various ways during its life cycle [5]. For example, it has been shown that young coffee trees suffer competition with different weed species under both controlled conditions [12–16] and in field studies [6, 17, 18]. Reduced plant growth has correlated with decreased photosynthetic efficiency [19] and nutrient accumulation by the branch [16, 20] and root systems of coffee plants [13] These studies also showed that the effect of weed competition on coffee was strongly dependent on both the weed species and density, and the age of the coffee plant after transplanting.

In another study, Ronchi et al. [20] verified severe competition in the relative content of macro (N, P, K, Ca, Mg, and S) and micronutrients (Zn, Cu, B, Fe, and Mn) in the aerial part of coffee plants when in competition with beggarticks (*Bidens pilosa*), dayflower (*Commelina diffusa*), motherwort (*Leonurus sibiricus*), apple-of-Peru (*Nicandra physalodes*), pusley (*Richardia brasiliensis*), and arrowleaf (*Sida rhombifolia*).

Therefore, IWM in coffee should consider the characteristics of individual weed species as well as their high nutrient recycling potential. Impaired crop growth due to weed competition soon after field transplanting will certainly cause irreversible losses in crop productivity [17].

3. Positive aspects of weeds

According to Souza et al. [21], weeds present in coffee plantations should be controlled to avoid loss of production and to facilitate farming and harvesting operations. On the other hand, if well managed, they can be beneficial to the crop, by contributing to shading the soil, avoiding direct sunlight (shading soil); mitigating the effects of erosion during the period of greater rainfall; and increasing the organic matter content of the soil through the decomposition of roots and aerial parts. However, it is important to avoid the production of weed seeds.

4. Common weeds in coffee plantations

The practice of surveying the predominant weed population in the cultivation area is considered of great importance, identifying its species and knowing its main characteristics, in order to support decision-making for the most appropriate control. The composition of the floristic community is always subject to the occurrence of variations, influenced by regional conditions, soil characteristics, type of exploration, and management system, which contribute to a greater or lesser presence of certain species in a given place and period. In coffee growing, we can group the main predominant weed species, highlighting the classifications as to the period of occurrence (dry and rainy), life cycle (annual and perennial), and type of leaf (narrow and

Monocotyledons					
Common name	**Latin name**	**Family**	**Infestation period**	**Cycle**	**Leaf**
Dayflower	*Commelina benghalensis*	Commelinaceae	Dry and rainy	Perennial	Broad
Nutsedge	*Cyperus rotundus*	Cyperaceae	Dry and rainy	Perennial	Narrow
Brazilian satintail	*Imperata brasiliensis*	Poaceae	Dry and rainy	Perennial	Narrow
Bermudagrass	*Cynodon dactylon*	Poaceae	Dry and rainy	Perennial	Narrow
Bahiagrass	*Paspalum notatum*	Poaceae	Dry and rainy	Perennial	Narrow

Monocotyledons					
Common name	**Latin name**	**Family**	**Infestation period**	**Cycle**	**Leaf**
Kikuyu grass	*Penisetum clandestinum*	Poaceae	Dry and rainy	Perennial	Narrow
Jamaican crabgrass	*Digitaria horizontalis*	Poaceae	Rainy	Annual	Narrow
Alexandergrass	*Urochloa plantaginea*	Poaceae	Rainy	Annual	Narrow
Sandbur	*Cenchrus echinatus*	Poaceae	Rainy	Annual	Narrow
Jaraguagrass	*Hyparrhenia rufa*	Poaceae	Dry and rainy	Perennial	Narrow
Guineagrass	*Panicum maximum*	Poaceae	Dry and rainy	Perennial	Narrow
Goosegrass	*Eleusine indica*	Poaceae	Rainy	Annual	Narrow
Sourgrass	*Digitaria insularis*	Poaceae	Dry and rainy	Perennial	Narrow
Dicotyledons					
Morningglory	*Ipomoea acuminata*	Convolvulaceae	Dry and rainy	Annual	Broad
Purslane	*Portulaca oleracea*	Portulacaceae	Dry and rainy	Annual	Broad
Radish	*Raphanus raphanistrum*	Cruciferae	Dry	Annual	Broad
Indigo	*Indigofera hirsuta*	Leguminosae	Dry and rainy	Perennial	Broad
Arrowleaf	*Sida rhombifolia*	Malvaceae	Dry and rainy	Perennial	Broad
Sanguinaria	*Alternanthera tenella*	Amaranthaceae	Dry and rainy	Perennial	Broad
Pigweed	*Amaranthus* spp.	Amaranthaceae	Dry and rainy	Annual	Broad
Buttonweed	*Borreria alata*	Rubiaceae	Dry and rainy	Annual	Broad
Pusley	*Richardia brasiliensis*	Rubiaceae	Dry and rainy	Annual	Broad
Poinsettia	*Euphorbia heterophylla*	Euphorbiaceae	Dry and rainy	Annual	Broad
Beggarticks	*Bidens pilosa*	Asteraceae	Dry	Annual	Broad
Sowthistle	*Sonchus oleraceus*	Asteraceae	Dry	Annual	Broad
Ageratum	*Ageratum conyzoides*	Asteraceae	Dry and rainy	Annual	Broad
Tasselflower	*Emilia sonchifolia*	Asteraceae	Dry	Annual	Broad
Marigold	*Tagetes minuta*	Asteraceae	Dry and rainy	Annual	Narrow

Dicotyledons					
Common name	**Latin name**	**Family**	**Infestation period**	**Cycle**	**Leaf**
Smallflower	*Galinsoga parviflora*	Asteraceae	Dry and rainy	Annual	Broad
Starbur	*Acanthospermum hispidum*	Asteraceae	Dry and rainy	Annual	Broad
Fleabane	*Conyza bonariensis*	Asteraceae	Dry and rainy	Annual	Broad

Source: Santos [22].

Table 1.
Main weed species prevalent in coffee plantations.

broad), consolidated in **Table 1**, according to Moraes et al. [23], Souza et al. [24], IBC [25], Silveira et al. [26], Matiello [27], and Matiello et al. [28].

5. Weed control methods

In weed control, besides chemical control, there are other methods that are efficient, economical, and beneficial to the coffee plant and to the environment that can be used on any farm. The management of weeds for sustainable agriculture is partitioned into (a) preventive management, (b) cultural management, (c) biological control, (d) physical control, (e) mechanical control, and (f) chemical control (herbicide).

5.1 Preventive management

Similar to cultural methods, preventive management for weed suppression are low-cost and advantageous for the coffee crop. According to Ronchi and Silva [5], there are very few but relatively important preventive methods that should be applied in coffee production systems, either to curb the entry or to decrease the dispersion of weed seeds in coffee plantations, they follow below:

- Care for seeds in soil correctives (straw and manure).

- Keeping farm roads free of weeds by clearing them or applying herbicides.

- Cleaning machinery during or after any mechanized operation on the farm.

- Remove any new weed infestations before they become more dense.

- Controlling weed species until the flowering stage to prevent seeds from spreading through the area by mechanical operations and animals, or to avoid increasing the weed seed bank in the soil [29].

- In areas of Mechanized Harvesting, the cleaning of the harvester should be performed. According to Matiello et al. [9], mechanized coffee harvesting has contributed to the dispersion of morningglory (*Ipomoea* spp.) seeds in crops, and this species should be controlled in its initial stage of development or by cleaning harvesters frequently to prevent infestation.

5.2 Cultural management

In coffee plantations in formation, a strip of 40–50 cm on each side of the planting line is kept free of weeds. In this case, the soil is exposed to solar radiation, the impact of rain, and the action of winds, all of which are harmful to the coffee plant, due to water evaporation and excessive heating of the first 10 cm of the soil surface. Currently, many producers work with intercropping between coffee trees and Congo grass (*Urochloa ruziziensis*) and signal grass (*Urochloa decumbens*). In this intercropping, the forage is cultivated between the rows (**Figure 1**), while the coffee planting row is kept covered by the residue thrown by the mower, during the mowing between the rows.

In soil exposed to the sun, plant growth is impaired by soil temperature and also by the evaporation of up to 15,000 liters of water per hectare per day [30]. The deposition of 5 t ha^{-1} of mown palisade grass (*Urochloa brizantha*) biomass, on the street of the coffee plantation, provides the equivalent of 70 kg ha^{-1} of nitrogen (N) and 8 kg ha^{-1} of potassium (K_2O). In a palisade grass pasture cultivated for 10 years without fertilizers, 45% more available phosphorus was found in soil samples taken under the clumps, compared to samples between the clumps [31].

Figure 1.
Consortium of Congo grass (Urochloa ruziziensis) with coffee, Larga farm, Ibiá, MG, Brazil. Photo: Daniel Resende Fontes.

Cutting green manures, such as pinto peanut, slender leaf rattlebox, jack bean, velvet bean, and millet, forms over time a layer of mulch that protects the soil and prevents or hinders the germination of the seeds of photoblastic positive weeds [32], which need light for their germination, Some examples of these weeds are: *Sida cordifolia*, *Sida rhombifolia*, and *Sida spinosa* [33] *Amaranthus* spp. and *Conyza* spp.

Millet is an annual grass (Poaceae) of tropical climate that has good resistance to drought, wide adaptation, and good mass production, in addition to fast growth, vigorous roots, and good capacity for nutrient cycling [34], considered a classic example of a cover crop, because it has a C/N ratio of 30 or higher in the budbreak and flowering phases [35], and can be an interesting option for cultural management and green manure.

Partinelli et al. [36], studying the effects of control treatments (no planting of cover crops), millet and the legumes pigeon pea, velvet bean, and cowpea, found that the biological fixation of nitrogen contributed about 80% of N accumulated by legumes, and depending on the production of dry biomass the contribution ranged from 27 to 35 kg N ha^{-1}. The pigeon pea (29.1 g kg^{-1}) and velvet bean (32.6 g kg^{-1}) showed the highest concentration of N.

On the other hand, regarding coffee plantations in formation, in organic and conventional systems, it was found that the bean straw mulch formed a physical barrier against weeds, providing soil coverage in the control of coffee weeds, obtaining satisfactory control and retaining more moisture in the soil, besides enabling the process of mineralization of this straw, which benefits the coffee in the organic system [37].

There are studies that have shown that residues of coffee husk and leaves caused inhibition of the germination of several wild species such as *Amaranthus retroflexus*, *Bidens pilosa*, *Cenchrus echinatus*, and *Amaranthus spinosus*, because of the release of allelopathic substances [38].

Martins et al. [39] found that plots subjected to *Mucuna deeringiana* mulch between the rows showed more than 90% reduction in weed density that was attributed to the allelopathic effects of this mulch.

In fact, different types of organic materials, including coffee waste such as coffee pulp, husk [40], and beans [41], have the potential to be used to control weeds through cover crop applications. For example, Yamane et al. [41] recently demonstrated that cover application of coffee grounds at 16 kg m^{-2} resulted in significant weed control for half a year. This inhibition was a result of an allelopathic effect due to the presence of caffeine, tannins, and polyphenols in coffee grounds [42].

Knowledge of the specificity of the allelopathic potential of plant residues will allow the efficient use of this resource in coffee growing as a practice in conventional coffee production, and especially in the production of certified coffee, whose products have a niche market with great prospects for expanding international demand.

Based on this information, we conclude that keeping the coffee trees permanently clean in the skirt area (chemical control) and with the weeds between the rows controlled by a rotary weeder (mechanical control) has stood out as a method that has maintained the principles of sustainability [43], besides producing organic matter for the coffee trees.

5.3 Biological control

The biological control method basically consists of using an agent that keeps the weed population at a lower level than would occur naturally, causing no economic damage to the crop.

The use of animals for weed control is hardly practiced anymore in modern coffee farming. This method consists of using ruminant animals (sheep) or birds (chickens) that will feed on the weed, thus reducing their population. The use of this method is little known in Brazilian coffee growing, and more investment in research is needed for it to become an alternative in the future.

5.4 Physical control

As emphasized in the sections above, if the weed vegetation is kept at a sufficient distance from the coffee row (to avoid resource competition), there is no need to eliminate the vegetation from the entire area (except during the harvest period in some countries) [5]. In addition, cover crops (mulching) or green manure can be successfully intercropped with coffee, as reported in the crop control.

Vegetable residues from other crops (if available on the farm at no additional cost), from the coffee tree (leaves and stems), or from tree branches, especially after pruning, can be used as mulching [5]. And the use of polyethylene plastic on the coffee row is also considered mulching.

5.5 Mechanical control

Manual weeding is one of the most important control methods on coffee farms, although they are slow and laborious [5]. During the formation stages, if preventive measures fail or if selective herbicides are not used, weeds that eventually germinate should be removed during the seedling formation and growth period [4]. Two years after field transplanting, several manual weeding operations are recommended to establish and maintain an adequate weed control range along the coffee rows, although herbicides can also be applied judiciously. On coffee farms where selective pre-emergence herbicide is applied as the primary method of weed control in the coffee rows, at least one manual weeding operation is performed 2–3 weeks after coffee planting, prior to herbicide application to regulate the soil surface and remove weeds.

The mechanical control of weeds is widely accepted by producers as a replacement or complement to other methods, especially manual ones, due to the fact that these methods have a higher yield, faster, and more economical. The difficulty of hiring labor, its high cost, and low yield, make the option for mechanical methods essential for large farms, being executed with the application of appropriate management techniques. These methods have great application in coffee farming, but they depend on the availability of equipment, spacing between rows, size of the plantation, slope index, and complementary methods of weed control. The most used implements coupled to tractors are the following:

- Grazer: normally with 2 knives, activated by the tractor's power takeoff, it is the most used implement in coffee farming, because it reduces the dissemination of weed seeds, being used at any time before flowering and fruiting, avoiding the formation of soil erosion processes. It must be used in the rainy and hot seasons of the year in coffee plantations with wider spacing. With adequate management, it is possible to keep weeds growing with controlled growth and to have the deposition of plant residues after cutting, forming mulch on the soil surface. In this operation some weed roots may die, which contributes to the formation of channels in the soil, favoring its aeration and water infiltration. Excessive use of the brush cutter can cause soil compaction, dominance of creeping weeds, and sprouting of some species, especially perennials.

- Brush: Contains a set of blades with a movement similar to that of a hammer mill, which grinds the weeds and plant residues such as branches and leaves. Several brands on the market with various types of blades and hammer, which presents greater efficiency over larger weeds and small tangled bushes, producing a thick layer of mulch over the soil.

5.6 Chemical control

The chemical method, or the use of herbicides, is a practice widely used in coffee farming, but for a better yield and effectiveness, the farmer must be careful in the correct choice of herbicide to be used in the field, according to several factors such as community, weed infestation level and stage of development, crop phase, soil type, time of application, toxicology of the herbicide, cost, equipment, and skilled labor in the application, in order to maximize efficiency while minimizing the effect on the environment [44].

Advantages:

- Speed and good operational yield
- Better applicability during rainy periods
- Keep the soil intact (without disturbance)
- Gradual weed disinfestation (perennial and vegetative propagated)
- Low cost and good efficiency in weed control

Disadvantages:

- Can cause injury, due to the drift effect
- When used in excess, can expose a lot of soil
- Requirement of adequate equipment with permanent maintenance
- Need for more training of producers or specialized labor
- May select resistant or tolerant weeds

Mixing herbicides is an important common practice to increase the spectrum of weed control in coffee plantations [5], compared to other crops, there are few herbicide formulations available for coffee. Herbicides are characterized by observing three main aspects [28]:

1. Selectivity: selective for the crop or non-selective (full action)
2. Season of use: used in PRE- or POST-emergence

3. Translocation in weeds: contact or systemic

These herbicides should be applied in a directed spray to the soil (PRE) or to weeds, respectively, to avoid injury to the coffee plant, for example, oxyfluorfen, is not completely selective on Arabica coffee [18] and to overcome the umbrella effects of higher coffee plants, the application doses of these herbicides should be determined based primarily on the physicochemical characteristics of the soil for herbicides applied PRE, and on herbicides in POST, the weed species and the stage of their development. On adult coffee plants, herbicides are mainly used between the rows, but applications in the coffee plant row may be necessary (e.g., to control *Ipomoea* spp.) [5]. In between the rows, herbicides have often been used during the rainy season for weed control in a narrow band beyond the projected skirt of the coffee plant. Total or partial desiccation in the strip, the weed residues are retained in the soil, contributing to soil and water conservation, nutrient cycling, and organic matter accumulation.

When recommending herbicides for the coffee crop, see **Table 2**, which consolidates the identification of the most commonly used herbicides, with their application times, dosages per hectare, and spectrums of action [28, 29, 46–50].

Chemical weed control in coffee farming became public through the replacement of the total-action, post-emergence, non-systemic, and highly toxic herbicide paraquat (banned in Brazil) by glyphosate, a systemic herbicide, also non-selective to coffee trees and applied post-emergence with low toxicity [51]. Due to its low cost, high

Application time	Commercial product	Active ingredient	Doses/ha	Narrowleaves	Action form
PRE	Goal and Galigan	Oxyfluorfen	2.0–6.0 L	Broad and narrow leaves	Contact
PRE	Sencor	Metribuzin	1.0–2.0 L	Broadleaves	Systemic
PRE	Alion	Indaziflam	0.15–0.20 L	Broad and narrow leaves	Systemic
PRE	Falcon	Pyroxasulfone + flumioxazin	0.45–1.0 L	Broad and narrow leaves	Contact and systemic
PRE	Boral and Stone	Sulfentrazone	1.4–2.0 L	Broad and narrow leaves	Systemic
PRE and POST	Flumyzin 500	Flumioxazin	0.05–0.240 L	Broadleaves	Contact
POST	Round up Original	Glyphosate	3.0–5.0 L	Broad and narrow leaves	Systemic
POST	Finale	Glufosinate-ammonium	2.0–3.0 L	Broad and narrow leaves	Contact
POST	Heat	Saflufenacil	35–100 g	Broadleaves	Contact
POST	Ally	Metsulfuron-methyl	6–10 g	Broadleaves	Systemic
POST	Aurora	Carfentrazone-ethyl	75–125 g	Broadleaves	Contact
POST	Clorimurom	Chlorimuron-ethyl	50–80 g	Broadleaves	Systemic
POST	Verdict Max	Haloxyfop-P-methyl	0.185–0.290 L	Narrowleaves	Systemic

Source: ADAPAR [45].

Table 2.
Main herbicides recommended for coffee plantations.

availability in the market, excellent toxicological profile and large number of controlled species, both grasses and broadleaves, the main herbicide used in coffee culture is glyphosate [52]. Repeated application during a season using the same active ingredient can select tolerant plants or resistant biotypes.

In order to control weeds of resistant biotypes, and avoid selection of new biotypes, herbicide associations are recommended for the control of a greater amount of weeds [53, 54]. The search for alternatives for the control of these resistant species, through IWM, find strategies to reduce the selection pressure of these biotypes such as reducing weed infestation, adopting an efficient green manure system, integrating and alternating control methods, such as preventive and cultural methods associated with chemical methods, alternating or associating herbicides with different mechanisms of action and using herbicides with different metabolism routes.

6. Integrated weed management (IWM)

IWM in coffee is based on the rational combination of different weed control practices (e.g., preventive, cultural, mechanical, biological, physical, and chemical) [5]. Every weed control system in coffee plantations should always be reviewed and analyzed with criteria every year, observing its effect on the soil and culture, as well as its technical and economic feasibility, respecting the conditions of each plantation [55]. Thus, no weed control practice is used in isolation [10].

Figure 2.
Integrated weed management (IWM) at Alquino farm, Pratinha, MG, Brazil. PRE-emergence herbicide application (A), mowing of Urochloa ruziziensis (B), mulching in the coffee row (C), and mulching in the coffee row (D). Photo: Daniel Resende Fontes.

Season	Time Brazilian	Crop	Weed control	
			In the intercrop rows	In the crop rows
Beginning of rains	September to November	Current	Planting of *Urochloa ruziziensis* (cultural control)	PRE and POST herbicide application
		Next	Mechanical grazer (throwing green matter on the coffee line) (mechanical control)	Green cover (cultural and biological control)
		Later	Mechanical grazer (throwing green matter on the coffee line) (mechanical control)	Green cover (cultural and biological control)
During the rains	December to February	Current Next Later	Mechanical grazer (throwing green matter on the coffee line) (mechanical control)	Green cover (cultural and biological control)
End of rains (tilling)	March to May	Current	POST herbicide application (chemical control)	No need (cultural control)
		Next	Blade carving (mechanical control)	
During the dry season (spraying)	June to August	Current	Push the mulch from the rows to the inter-row (furrower) (mechanical control)	Mulching (cultural control)
		Next		Straw application (biological and cultural control)

Source: Adapted from Santos [56].

Table 3.
Suggestion for integrated weed management (IWM) in coffee plantations.

Every weed control system in coffee plantations should always be reviewed and analyzed with criteria every year, observing its effect on soil and crop, as well as its technical and economic feasibility, respecting the conditions of each plantation [55]. The IWM of coffee consists of the union of all types of control (**Figure 2**), applied in a combined, successive, and rotational manner at a given time and space, considering the conditions of the plantation and the execution of other agricultural practices.

Priority should be given to carrying out different controls in order to take advantage of the available resources and achieve greater efficiency, reduce costs, and obtain maximum safety for humans and minimum damage to the environment (**Table 3**).

7. Conclusions

The combination of weed control methods has proven to be a sustainable practice in coffee production. In integrated weed management, the inherent advantages of each control method must be combined, considering requirements such as safe application, age, spacing, and size of the plantation, as well as full knowledge of the weeds, their growth stage, leaf type, frequency, and population density. By reinforcing the study of the biology and physiology of weeds, we can guarantee the formation of a consistent diagnosis, which will provide an efficient control action with lower costs, better environmental conservation, and higher crop productivity.

Acknowledgements

The authors are grateful to the "Fundação de Amparo à Pesquisa do Estado de Minas Gerais" (FAPEMIG-2070.01.0004768/2021-84) and "Conselho Nacional de Desenvolvimento Científico e Tecnológico" (CNPq – 404240/2021-6) for financial support.

Conflict of interest

The authors declare no conflict of interest.

Author details

Daniel Resende Fontes[1], Andrew de Paula Ribeiro[1], Marcelo Rodrigues dos Reis[2], Miriam Hiroko Inoue[3] and Kassio Ferreira Mendes[1*]

1 Department of Agronomy, Federal University of Viçosa, Viçosa, Minas Gerais, Brazil

2 Institute of Agricultural Sciences, Federal University of Viçosa, Viçosa, Minas Gerais, Brazil

3 Department of Agronomy, Mato Grosso State University, Tangará da Serra, Mato Grosso, Brazil

*Address all correspondence to: kfmendes@ufv.br

References

[1] IBGE – Instituto Brasileiro de Geografia e Estatistica. Café. 2019. Available from: https://www.ibge.gov.br. [Accessed June 2, 2022]

[2] CONAB – Companhia Nacional deAbastecimento. Acompanhamento da Safra Brasileira de Café – Safra 2021. n. 1. Brasília, DF: CONAB; 2021. p. 72

[3] Damatta FM et al. Coffee: Environment and crop physiology. In: Damatta FM, editor. Ecophysiology of Tropical Tree Crops. Vol. 3. New York: Nova Science Publishers; 2010. pp. 181-216

[4] Ronchi CP, Silva AA, Ferreira LR. Manejo de plantas daninhas em lavouras de café. Viçosa, MG, Brazil: UFV; 2001

[5] Ronchi CP, Silva AA. Sustainable weed control in coffee. In: Korres NE et al., editors. Weed Control Sustainability, Hazards and Risks in Cropping Systems Worldwide. Boca Raton, FL: CRC Press; 2018. pp. 425-441

[6] Araújo F, Ronchi CP, Almeida WL, et al. Optimizing the width of strip weeding in Arabica coffee in relation to crop age. Planta Daninha. 2012;**30**:129-138

[7] Alcântra EN, Ferreira MM. Efeitos de métodos de controle de plantas daninhas na cultura do cafeeiro (*Coffea arabica* L.) sobre a qualidade física do solo. Revista Brasileira de Ciência do Solo. 2000;**24**:711-721

[8] Schroth G, Laderach P, Dempewolf J, et al. Towards a climate change adaptation strategy for coffee communities and ecosystems in the Sierra Madre de Chiapas, Mexico. Mitigation and Adaptation Strategies for Global Change. 2009;**14**:605-625

[9] Matiello JB et al. Cultura de café no Brasil: Manual de recomendações, São Paulo, SP, Brazil: Futurama Editora, 2016

[10] Bajwa AA. Sustainable weed management in conservation agriculture. Crop Protection. 2014;**65**:105-113

[11] Oliveira JA, Matiello JB, Carvalho F. Estudo do efeito da época de controle das plantas daninhas sobre a produção do café. Congresso Brasileiro de Pesquisas Cafeeiras. 1979;7:350-352

[12] Ronchi CP, Silva AA. Effects of weed species competition on the growth of young coffee plants. Planta Daninha. 2006;**24**:415-423

[13] Ronchi CP, Terra AA, Silva AA. Growth and nutrient concentration in coffee root system under weed species competition. Planta Daninha. 2007;**25**:679-687

[14] Fialho CM, Silva GR, Freitas MAM, et al. Competition of weeds with coffee plants, in two times of infestation. Planta Daninha. 2010;**28**:969-978

[15] Fialho CMT, França AC, Tironi SP, Ronchi CP, Silva AA. Interferência de plantas daninhas sobre o crescimento inicial de *Coffea arabica*. Planta Daninha. 2011;**29**:137-147

[16] Carvalho LB, Alves PLCA, Duke SO. Hormesis with glyphosate depends on coffee growth stage. Anais da Academia Brasileira de Ciências. 2013;**85**:813-821

[17] Lemes LN et al. Weed interference on coffee fruit production during a four-year investigation after planting. African Journal of Agricultural Research. 2010;**5**:1138-1143

[18] Magalhães CEO, Ronchi CP, Ruas RAA, et al. Seletividade e controle

de plantas daninhas com oxyfluorfen e sulfentrazone na implantação de lavoura de café. Planta Daninhas. 2012;**30**:607-616

[19] Rossmann M, Matos AT, Abreu EC, Silva FF, Borges AC. Effect of influent aeration on removal of organic matter from coffee processing wastewater in constructed wetlands. Journal of Environmental Management. 2013;**128**:912-919

[20] Ronchi CP, Terra AA, Silva AA, Ferreira LR. Acúmulo de nutrientes pelo cafeeiro sob interferência de plantas daninhas. Planta Daninha. 2003;**21**:219-227

[21] Souza IF et al. Plantas daninhas e seu controle. Informe Agropecuário. 1985;**11**:59-65

[22] Santos JCF et al. Manejo Integrado das Plantas Infestantes no Cafezal. Circular Técnica, 69, EMBRAPA, 2004

[23] Moraes FRP. Práticas de cultivo. In: Graner EA, Junior CD, editors. Manual do cafeicultor. São Paulo, SP, Brazil: Editora Melhoramentos; 1967. pp. 127-151

[24] Souza IF et al. Controle de ervas daninhas. Informe Agropecuário. 1978;**4**:56-66

[25] IBC (Rio de Janeiro, RJ). Cultura do café no Brasil: pequeno manual de recomendações. Rio de Janeiro: IBC/DIPRO; 1986

[26] Silveira CA et al. A comprovada eficiência de sencor. Correio Agrícola. 1988;**1**:8-10

[27] Matiello JB. O Café: do cultivo ao consumo. Porto Alegre, RS, Brazil: Editora Globo; 1991

[28] Matiello JB et al. Controle do Mato em Cafezais. Varginha, MG, Brazil: SARC/PROCAFÉ; 2009

[29] Ronchi CP et al. Manejo de plantas daninhas na cultura do café. In: Monquero PA, editor. Manejo de plantas daninhas nas culturas agrícolas. São Carlos, SP, Brazil: RiMa Editora; 2014. pp. 132-154

[30] Ragassi CF et al. Aspectos positivos e riscos no consorcio cafeeiro e braquiária. Visão Agrícola. 2013:12

[31] Corazza EJ et al. Spatial variability of soil phosphorus of a low productivity *Brachiaria brizantha* pasture. Scientia Agricola. 2003;**60**:559-564

[32] Santos IC et al. Manejo de entrelinhas em cafezais orgânicos. Informe Agropecuário, Belo Horizonte. 2002;**23**:115-126

[33] Felipe GM, Polo M. Germinação de ervas invasoras: efeito da luz escarificação. Revista Brasileira de Botânica. 1983;**6**:55-60

[34] da Silva RH, Rosolem CA. Early development and nutrition of cover crop species as affected by soil compaction. Journal of Plant Nutrition. 2003;**26**:1635-1648

[35] Kliemann HJ, Braz AJB, Silveira PM. Taxas de decomposição de resíduos de espécies de cobertura em Latossolo Vermelho distroférrico. Pesquisa Agropecuária Tropical. 2006;**36**:21-28

[36] Partelli FL, Vieira HD, Espindola JAA, Urquiaga S, Fernandes EP, Pacheco LP. Fixação biológica de nitrogênio por plantas de cobertura cultivadas na entrelinha de cafeeiro Conilon orgânico. In: VI SIMPÓSIO DE PESQUISA DOS CAFÉS DO BRASIL. Vitória, 2009. Brasília/D.F.: Embrapa - Café

[37] Cunha RL et al. Desenvolvimento e produtividade do cafeeiro orgânico. Simpósio de Pesquisa dos Cafés do Brasil. 2003:406-407

[38] Almeida FS. Efeitos alelopáticos de resíduos vegetais. Pesquisa Agropecuária Brasileira. 1991;**26**:221-236

[39] Martins BH et al. Soil organic matter quality and weed diversity in coffee plantation area submitted to weed control and cover crops management. Soil and Tillage Research. 2015;**153**:169-174

[40] Minassa EMC et al. Efeito alelopático da palha de café (*Coffea canephora* L. e *Coffea arabica* L.) sobre plantas cultivadas e espontâneas [Doutorado em Produção Vegetal, Tese]. 91 f. 2014. Universidade Federal do Norte Fluminense, Campos dos Goytacazes, RJ

[41] Yamake KM et al. Field evaluation of coffee grounds application for crop growth enhancement, weed control, and soil improvement. Plant Production. 2014;**17**:93-102

[42] Pandey A et al. Biotechnological potential of coffee pulp and coffee husk for bioprocesses. Biochemical Engineering Journal. 2000;**6**:153-162

[43] Alcântra EN. Efeito de diferentes métodos de controle de plantas daninhas na cultura do cafeeiro (*Coffea arabica* L.) sobre a qualidade de um Latossolo Roxo distrófico. MG, Brazil: Universidade Federal de Lavras; 1997

[44] Blanco, FMG. Controle Químico das plantas Daninhas na Cultura do Café. Biológico, vol. 1 2004;**138**:147

[45] ADAPAR (Agência Agropecuaria de Defesa do Paraná). Sistema de Agrotóxicos Fitossanitários, 2011. Available from: https://www.adapar.pr.gov.br. [Accessed September 10, 2022]

[46] Aguiar V, Staver C, Milberg P. Weed vegetation response to chemical and manual selective ground cover management in a shaded coffee plantation. Weed Research. 2003;**43**:68-75

[47] Sánchez L, Gamboa E. Control de malezas con herbicidas y métodos mecánicos en plantaciones jóvenes de café. Bioagro. 2004;**16**:1-4

[48] Gómez R. Efecto del control de malezas con paraquat y glifosato sobre la erosión y pérdida de nutrimentos del suelo en cafeto. Agronomía Mesoamericana. 2012;**16**:77-87

[49] Matiello JB et al. Cultura do Café no Brasil: Manual de Recomendações. Varginha, MG, Brazil: SARC/PROCAFÉ; 2011

[50] Rodrigues BN, Almeida FS. Guia de herbicidas. 7th ed. Londrina, PR, Brazil: Edição dos Autores; 2018. p. 764

[51] Alcântra EN, Silva RA. Manejo do Mato em Cafezais. In: Reis PR, Cunha RL (Eds). Café arábica do plantio a colheita. Lavras: EPAMIG. 2010;**1**:519:572

[52] Christoffoleti PJ, Nicolai M. Convivência com plantas daninhas não deve limitar cafezal. Visão Agrícola. 2013;**12**:3

[53] Mendes KF, Silva AA. Applied Weed and Herbicide Science. 1st ed. Cham, Switzerland: Springer; 2022. p. 299

[54] Silva AA et al. Manejo Integrado de Plantas Daninhas. In: Sakiyama NH et al., editors. Café arábica: do plantio à colheita. Viçosa: UFV; 2015. pp. 104-128

[55] Alcântra EN et al. O manejo do mato em cafeeiros. Informe Agropecuário. 1989;**14**:2-28

[56] Santos IC et al. Manejo de plantas daninhas no cafezal. Boletim Técnico EPAMIG. 2000;**61**:24

Chapter 6

Toxicological Interaction Effects of Herbicides and the Environmental Pollutants on Aquatic Organisms

Mahdi Banaee

Abstract

Although herbicides are designed to remove or control weeds, pollution of water ecosystems with herbicides could have adverse effects on aquatic animals such as fish. The effect of herbicides on nontarget organisms may be different than expected, as herbicides may interact with another environmental contaminant. Since there are different contaminants in the water, fish may live in the cocktail of xenobiotics, including herbicides. Therefore, herbicides alone and in combination with other pollutants could affect fish physiology. Thus, the interaction of environmental contaminants with pesticides may create a situation in which a chemical affects the activity of a pesticide; that is, its effects increase or decrease or produce a new effect that neither of them creates on its own. These interactions may occur due to accidental misuse or lack of knowledge about the active ingredients in the relevant materials. This study aimed to review the effects of herbicides alone and in combination with other xenobiotics on various aspects of fish biology. In this study, different biomarkers were reviewed in fish exposed to herbicides.

Keywords: biomarkers, herbicides, aquatic ecosystems, xenobiotic, aquatic animals

1. Introduction

The agricultural revolution is the starting point for using various types of pesticides and synthetic and chemical fertilizers to increase agriculture crops' volume and maintenance [1–3]. Thus, the development agriculture industry has caused an increase in the pollution of aquatic ecosystems with agrochemicals. Pesticides, including herbicides, are pollutants that can be found in the water around agriculture fields. Herbicides are usually used to control weeds and unwanted plants in agriculture farms, fruit gardens, aquaculture ponds, and urban green spaces [3, 4]. Herbicides may enter water ecosystems when used or after being applied. Penetrating herbicides into surface and groundwaters may occur through the drainage of agriculture farms during spraying or after that [5]. Although herbicides may enter water bodies through the drainage of agricultural fields, they can also be used to control weeds in pools or lagoons. Therefore, they can affect water ecosystems directly or indirectly [4].

Studies showed that herbicides could be detected in the drinking water. For example, concentrations of glyphosate in drinking water in the United States and Australia were 700 μg L^{-1} and 1000 μg L^{-1}, respectively [6].

Tracing some herbicides, such as atrazine, acetochlor, and 2,4-D, in groundwater [7], streams [8], river [9], lake [10], marine ecosystems [11], and estuaries [12] indicates that herbicides are highly mobile. Toxicological data showed that more than 99% of pesticides never affect target organisms. In other words, a significant part of pesticides is released into the environment and influences nontarget organisms [13]. Therefore, the different concentrations of herbicides can impact aquatic organisms' health. Similar reports indicate that even humans and pets are exposed to herbicides.

Although herbicides' chemical structure is designed to affect weeds, they could have toxicity effects on aquatic animals. Herbicides are lipophilic compounds that can easily cross biological barriers and penetrate animals' bodies. The physiological and behavioral changes in aquatic animals exposed to herbicides indicate that herbicides have a potentially toxic effect on nontarget animals. We could observe toxicity effects after aquatic organisms' exposure to herbicides.

Herbicides may be absorbed via gills, skin, or intestinal epithelium. Next, they may enter the blood and distribute it in the various tissues by circulating blood. Although herbicides may be repelled in the urine and feces, they may be reached into the liver via the blood circulation system and metabolized in the hepatocytes by detoxification enzymes. A significant part of herbicides may conjugate with a nonenzyme antioxidant such as glutathione and excrete quickly. Other part of metabolites may be repelled through renal and digestive systems; however, reactive oxygen species (ROS) and some metabolites produced during detoxification remain in animals' bodies. These metabolites and ROS may be conjugated with nonenzyme antioxidants and removed or may be neutralized by antioxidant enzymes. Reactive oxygen species production in the detoxification process of herbicides can induce oxidative stress in aquatic organisms. This phenomenon would occur if detoxification mechanisms in the liver work very well or animals are exposed to a sublethal dose of herbicides. Otherwise, various toxicity effects would be detected in organisms challenged by herbicides.

This chapter aims to illustrate toxicology herbicides to fill gaps in information about the toxicity effects of herbicides on aquatic animals. In this chapter, we try to provide documentation on the effects of herbicides on various aspect of aquatic animals' biology. In addition, we will discuss the interaction of other xenobiotics with herbicides.

2. Interaction of herbicides with other xenobiotics

The natural aquatic ecosystems usually contain various xenobiotics that can affect fish [14, 15]. In other words, fish may live in the cocktails of different pollutants [16, 17]. Thus, fish must be able to survive and resist a range of environmental pollutants [18].

Furthermore, various contaminants may interact with each other [19, 20]. Interaction between pollutants includes additive effects and synergic or antagonistic effects. In the additive and synergistic effects, toxicity and bioavailability of xenobiotics are increased. In contrast, in the antagonistic situation, one or more pollutants reduce toxicity and bioavailability of other xenobiotics [21, 22].

Tabche, et al. [23] studied the combined effects of paraquat and lead (Pb) on the liver of *Oreochromis hornorum*. They found that paraquat and lead had synergistic effects on fish. A synergic effect of microplastic on paraquat toxicity was shown in common carp (*Cyprinus carpio*) by Nematdoost Haghi and Banaee [22]. Also, Xu, et al. [24] displayed that exposure of goldfish (*Carassius auratus*) to paraquat and Pb caused activation of detoxification enzymes in the hepatocytes. The effect of iron oxide nanoparticles (γ-Fe2O3) and glyphosate on the liver of *Poecilia reticulata* was assayed by de Lima Faria, et al. [25]. Changes in the biochemical parameters were detected in the crayfish (*Astacus leptodactylus*) exposed to glyphosate and chlorpyrifos [26, 27]. Bonifacio, Zambrano and Hued [28] displayed that co-exposure to glyphosate and chlorpyrifos changed blood biochemical parameters in *Cnesterodon decemmaculatus*.

3. Biological response of aquatic organisms to herbicides

Therefore, to understand the herbicide effects on aquatic life, herbicide's anecdote is told since primarily its entered aquatic ecosystems, in this chapter. Then, it is said about herbicide's fate in animal's body to its excretion.

After draining herbicides in water ecosystems, they could penetrate the cellular membrane and cytoplasm. These chemical toxicants may influence cell permeability, ion transport, electron transport, and enzyme activities associated membrane. Next, herbicides could disrupt the cellular organelles' functions, which may lead to induce apoptosis, cell necrosis, or activation of the tumorigenesis in cells. Thus, herbicides could affect different functions of the biological membrane.

But the question that may be on readers' minds is whether animal cells are defenseless against herbicides? No!

4. Detoxification and metabolism of herbicides

In two phases, herbicides may be converted into excretable metabolites in hepatocytes of aquatic animals. Maternal compounds combine with oxygen and oxidize in the primary phase (Phase I), known as the biotransformation step. Then, oxidized metabolites are conjugated with water-soluble polar biomolecules in the cell (Phase II). Next, herbicides' metabolites may be excreted through urine or bile [29].

Active compounds as reactive oxygen species are often produced during detoxification that could cause the oxidation of macromolecules. However, a cellular antioxidant defense system could neutralize reactive oxygen species (ROS) and inhibit peroxidation reactions. There is a balance between ROS and cellular antioxidant defense capacity in normal conditions. If this balance is collapsed and ROS levels are more than cellular antioxidant defense potential, oxidative stress would occur. ROS attacks macromolecules in this situation, leading to severe histopathological damage to vital tissues.

The disruption in the detoxification enzymes' function may occur in the fish exposed to herbicides. Therefore, defects in the function of the detoxification system can make fish vulnerable to the toxicity of herbicides. A significant decrease in mitochondrial cytochrome content was reported in *Oreochromis niloticus* exposed to pendimethalin [30]. Zhang et al. [31] assayed mitochondria-immune responses in zebrafish, *Danio rerio* following challenge with dinoseb. They reported a significant decrease in the expression of genes involved in mitochondrial respiration and cellular detoxification [31].

We know very well that exposure of fish to xenobiotics such as herbicides could cause an imbalance between ROS contents and cellular antioxidant defense capacity [32]. Therefore, exposure of fish to herbicides could lead to oxidative stress. Damage to membrane phospholipids decreases the cellular chance of survival and increases apoptosis and necrosis rates. Disruption in the cellular membrane's physiological function also affects metabolism, biochemical hemostasis, gene expression, and DNA replication in the cells [15]. In the following, we want to explain the effects of herbicides on aquatic animals in more detail.

Involvement of cellular detoxification and biotransformation systems to remove xenobiotics may reduce its ability to detoxify herbicides. Therefore, the toxic effects of herbicides on fish would be increased if the detoxification mechanism was collapsed.

5. Oxidative stress

The oxidative stress in fish exposed to herbicides can be attributed to ROS. Furthermore, ROS production during the detoxification of other xenobiotics may further contribute to oxidative stress due to herbicide exposure.

Like other vertebrates, the antioxidant defense system of fish includes antioxidant enzymes and nonenzyme antioxidants. Therefore, change in the antioxidant enzyme activities and nonenzyme antioxidant contents are biomarkers that show activation of the antioxidant defense system against ROS. Pereira, Fernandes and Martinez [33] showed that hepatic antioxidant enzymes activated after exposure of *Prochilodus lineatus* to clomazone. Oxidative damage was seen in the hepatocytes of *O. niloticus* and *Geophagus brasiliensis* after treatment with mesotrione herbicide [34].

Changes in the antioxidant enzyme activities indicated oxidative stress in the gills and liver of tetra fish (*Astyanax altiparanae*) exposed to atrazine [35]. Moraes, et al. [36] found that oxidative stress occurred in the teleost fish (*Leporinus obtusidens*) after exposure to clomazone and propanil.

Otherwise, interaction of ROS with vital macromolecules such as DNA, lipids, proteins, etc., can lead to their peroxidation. Thus, these macromolecules may be lost their biological functions, and their metabolites may disrupt the cellular hemostasis.

In the assessment of oxidative damages, a measure of malondialdehyde, protein carbonyl, oxidized thiol groups, and 7,8-dihydro-8-oxoguanine (8-oxo-dG) is routine.

Malondialdehyde is a more critical metabolite produced during lipid peroxidation. Therefore, a significant increase in malondialdehyde contents in the target cells indicates oxidative stress. Moreover, an increase in the malondialdehyde expedites cascading reactions of lipid peroxidation. Protein carbonyl is known as a metabolite of protein oxidation. Furthermore, increasing the peroxidation rate of thiol groups can be a physiological response to ROS increase at the cellular level. A significant decrease in the total antioxidant and increase in the protein carbonyls and malondialdehyde contents were reported in the liver and brain of hybrid surubim (*Pseudoplatystoma* sp) exposed to glyphosate and roundup [37].

Also, a significant increase in 7,8-dihydro-8-oxoguanine (8-oxo-dG) contents is a biomarker of nucleic acid oxidation and gene damage.

However, other biomarkers can be used to detect oxidative stress indirectly. We will describe each of them in the following sections.

6. Neurotoxicity

Studies showed that xenobiotics could often influence nerve systems. Therefore, this is a problem in distinguishing the primary neurotoxicity agent in fish when exposed to herbicides combined with other pollutants. Thus, if we observed neurotoxicity response in fish, evaluation of the additive or synergistic effects of xenobiotics on herbicides' toxicity should be a priority.

Peroxidation of phospholipids that cover nerves can disrupt transport of neural signals or information processing in neural centers. Also, herbicides can change neurotransmitters' biochemical structure or disable enzymes involved in biosynthesis or biodegradation of neurotransmitters.

Moraes, et al. [36] found that exposure of teleost fish (*L. obtusidens*) to clomazone and quinclorac decreased acetylcholinesterase (AChE) activity in the brain, while AChE activity increased in muscle tissue after exposure to clomazone, propanil, and metsulfuron-methyl. Similarly, the inhibition of AChE activity was reported in the brain of teleost fish (*L. obtusidens*) exposed to herbicides clomazone and propanil [36]. Thanomsit et al. [38] could design a monoclonal antibody-ACHE that is used to detect acetylcholinesterase activity in the brain of fish exposed to herbicides. Thus, they could measure AChE activity in the brain of hybrid catfish, Nile tilapia, and climbing perch [38].

One of the consequences of neurotoxicity is the occurrence of behavioral changes in aquatic animals exposed to herbicides.

7. Behavioral response

Changes in the behavior of animals may be related to disrupting nerve systems or muscle spasms. Previous research showed that exposure to aquatic animals to herbicides could alter the behavior and rate of their response to environmental stimuli. Herbicides can affect the relationship between hunters and prey. Also, exposure to animals to herbicides may change animals' romantic, reproductive, and parenting behaviors. Thus, changes in feeding behavior can decrease the growth performance of organisms exposed to herbicides [39].

Faria et al. [25] documented that changes in the behavior of fish exposed to herbicides had a significant relationship with changes in the monoaminergic neurotransmitters in the brain. They found that a significant increase in dopamine (DA), serotonin (5-HT), and a decrease in norepinephrine (NE) could change the exploratory and social behaviors of zebrafish following exposure to glyphosate.

Butyrylcholinesterase (BChE) is known as pseudocholinesterase. Fluctuations in the BChE activity may change the behavior of aquatic animals. A significant change in the BChE activity was observed in freshwater fish *Labeo rohita* exposed to Roundup® [40]. Geetha [40] found that increased BChE activity could relieve the Roundup® induced stress in fish.

8. Genotoxicity and gene damage

The genotoxicity effects of herbicides may be due to the interaction of ROS with DNA [41]. Exposure to herbicides and their metabolites may degrade DNA or adduct to DNA structure. The DNA damage to erythrocytes, liver, and gills was detected by comet assay in the *O. niloticus* and *G. brasiliensis* exposed to Mesotrione [34]. DNA damage was reported in the European eel (*Anguilla anguilla*) exposed to Roundup® (glyphosate-based) and Garlon® (triclopyr-based) [42]. Ruiz de Arcaute, Soloneski and Larramendy [41] observed that exposure of *C. decemmaculatus* to dicamba could cause micronuclei and DNA single-strand breaks in circulating blood cells. Similar results were observed in the *P. lineatus* [43], *C. auratus* [44], and *C. decemmaculatus* [45] exposed to Roundup, atrazine, and glyphosate, respectively. DNA damage and genotoxicity were detected in the egg of silver catfish (*Rhamdia quelen*) exposed to 2,4-D and glyphosate [46].

Enhancement or depression in the mRNA expression of enzymes involved in detoxification and biotransformation of xenobiotics was reported in fish exposed to herbicides. For example, Velki, et al. [47] reported a significant increase in *Ces2* gene expression in the zebrafish embryos following the exposure to 2.15 μM diuron for 96 h. Exposure to Roundup and other glyphosate changed gene expression patterns in the reproductive tissue of Japanese medaka fish (*Oryzias latipes*) [48].

Increased genetic defects and neoplasia in fish embryos and larvae can be caused by exposure to xenobiotics [49], including herbicides. Also, mutation due to exposure of fish to herbicides may lead to tumor generation.

9. Blood biochemical parameters

Moreover, the rapture of cellular membranes may cause the release of cytoplasmic contents or organelles into intercellular fluid such as serum. Hence, assessing biochemical parameters in serum can indicate the stability of cellular membranes after exposure to herbicides [32]. Geetha [40] demonstrated that exposure to Roundup® could affect the balance of plasma electrolytes and transaminase activity in *L. rohita* [40]. The disruption in biochemical hemostasis was reported in the crayfish exposed to glyphosate and chlorpyrifos [26, 27].

The increase in the serum enzyme activities and changes in the blood biochemical parameters were observed in *C. carpio* exposed to paraquat [22]. Similar results were detected in *C. carpio* following glyphosate [50]. A significant change in glucose, cholesterol, and triglyceride levels in the blood may be due to elevated energy needs to alleviate the cytotoxic effects of herbicides.

10. Suppression of the immune system

Exposure to xenobiotics can suppress immune system functions by increasing corticosteroid hormones. A significant increase in corticosteroid hormones can affect cytokine gene expression. Thus, an increase in inflammation response can depress immune system power.

Maddalon, et al. [51] showed that glyphosate herbicide could induce immunotoxicity by interfering with the hormonal pathway and biosynthesis of cytokines and neuropeptides. Also, Acar, et al. [52] displayed that changes in the immune-related genes could mitigate immune functions in Nile tilapia (*O. niloticus*) exposed to glyphosate.

11. Reproductive disorders

Some herbicides can disrupt reproduction physiology. Herbicides may act as endocrine disruptors. They can block hormone receptors or induce changes in enzyme function involved in hormones' biosynthesis. Furthermore, some herbicides may act as analogs of natural hormones. Reproduction products may be denatured after animals' exposure to herbicides. Therefore, the rate of fecundity, fertility, and survival of embryos may be collapsed. This phenomenon can also affect the hatchling rate and percentage of larvae survival. Decreased adaptability of larvae to environmental conditions may be the reason for the reduced survival rate after exposure to herbicides [53].

Yusof, Ismail and Alias [54] found that exposure of Java medaka (*Oryzias javanicus*) to glyphosate reduced fertility, hatching eggs, and larval survival. Furthermore, Zebral, et al. [53] discovered that Roundup exposure changes the diapausing pattern of *Austrolebias nigrofasciatus* embryos. Thus, Roundup could affect the survival of *A. nigrofasciatus* embryos. Decreased fecundity rates were also observed in *A. nigrofasciatus* breeders exposed to Roundup. Also, Dehnert, Karasov and Wolman [55] displayed that 2,4-D exposure could reduce zebrafish and perch survival rates during larval stages. They explained that a decrease in the survival rate of larvae could be due to the toxicity effect of 2,4-D on the development and function of neural circuits underlying the vision of larval fish. Moreover, Dehnert et al. [56] revealed that the application of 2,4-D to control Eurasian watermilfoil (*Myriophyllum spicatum*) in aquatic ecosystems could threaten fish survival.

12. Growth dysfunction

Previous studies showed that herbicides could decrease growth performance in aquatic animals. A significant weight reduction may be related to disruption in nutrient absorption in digestive systems. Deficiency in the assimilation of vital macromolecules can alter energy budgeting. As a result, animals have to consume energy storage in the liver and muscles to supply their needs. Therefore, weight loss and general weakness, anorexia, were often reported in the aquatic animals exposed to herbicides [39].

13. Hemotoxicity

Herbicides could change white blood cell (WBC), red blood cell (RBC) counts, and hematological indexes such as hemoglobin and hematocrit contents in fish. These phenomena can be related to hematopoietic tissue damage. Moreover, disruption in the blood circulation systems may occur in fish exposed to herbicides. Hemolysis of erythrocytes, a decline in erythropoietin levels, and histopathological damage to hematopoietic organs can reduce blood cell counts in animals exposed to herbicides. Pereira, Fernandes and Martinez [33] declared that changes in the hematological parameters could be due to the toxicity effects of clomazone on the hematopoietic tissue of fish *P. lineatus*. Exposure of *P. lineatus* to clomazone changed hematological parameters after 96 h. Moreover, Merola, et al. [57] showed that exposure of zebrafish to pendimethalin could cause blood congestion, impair blood flow, and reduce heartbeat.

14. Histopathological damage

Histopathological injuries could be related to oxidative damage to the cellular membrane of fish exposed to herbicides. Furthermore, apoptosis and cellular necrosis may intensify histopathological damages in various tissues of fish exposed to herbicides. Destro, et al. [35] found that atrazine exposure could damage the liver tissue of tetra fish (*A. altiparanae*). They showed that the histopathological damage in the liver was due to an increase in lipid peroxidation. Moreover, Nassar, Abdel-Halim and Abbassy [30] reported histopathological damage in the gills and liver of fish exposed to the herbicide pendimethalin.

15. Bioaccumulation of herbicides

The bioaccumulation of xenobiotics is directly related to their bioavailability. Therefore, environmental pollutants that may increase the bioavailability of herbicides can significantly impact their bioaccumulation capacity in aquatic animals.

Furthermore, the half-life of herbicides in water ecosystems can also affect their bioaccumulation capacity. The half-life of herbicides in the various environments is different. Herbicides in environmental conditions can be quickly degraded into various metabolites. Some herbicides are durable in the environment. The breakdown rate of herbicides depends on their chemical structures and environmental conditions [20].

Therefore, the probability of their bioaccumulation in the body of aquatic animals is also high. Various authors reported the bioaccumulation of herbicides in aquatic animals. Tyohemba et al. [10] measured the bioaccumulation of various herbicides in African mud catfish (*Clarias gariepinus*), and Mozambique tilapia (*Oreochromis mossambicus*) inhabited Lake St. Lucia, South Africa. They detected phenoxy-acid herbicides, acetochlor, atrazine, and terbuthylazine in the muscle tissues of fish [10]. The analysis of fresh fish tissues collected from four markets in Nanning City, Guangxi Province, China, showed that the bioaccumulation of atrazine, acetochlor, metolachlor, and their metabolites could be worrying [58]. Furthermore, herbicides have also been found in fish and seafood [59, 60]. Therefore, the bioaccumulation of herbicides could threaten consumers' health.

16. Conclusion

We tried to present an overview of herbicides' toxicity in this chapter. However, we must update our information because newborn pollutants could be found in water ecosystems that can affect herbicides' half-life, toxicity, and bioavailability. Overall, if we want to discuss the effects of herbicides alone or in combination with other xenobiotics, we should be well known of their toxicity mechanisms and pathways and how they can affect the physiology of aquatic animals. Therefore, if we find the source of herbicide pollution, we can prevent their destructive effects on fish before penetrating aquatic ecosystems. Also, if we cognize about biotransformation and detoxification of herbicides, we can better manage the adverse effects of herbicides on fish. Therefore, studies on toxicity, bioavailability, and interaction of herbicides with other pollutants can be useful in recognizing the physiological response of fish exposed to herbicides.

Author details

Mahdi Banaee
Faculty of Natural Resources and the Environment, Aquaculture Department, Behbahan Khatam Alanbia University of Technology, Behbahan, Iran

*Address all correspondence to: mahdibanaee2@gmail.com

References

[1] Banaee M, Mirvagefei AR, Rafei GR, Majazi B, Amiri. Effect of sub-lethal diazinon concentrations on blood plasma biochemistry. International Journal of Environmental Research. 2008;**2**(2):189-198

[2] Banaee M, Sureda A, Mirvaghefi AR, Ahmadi K. Biochemical and histological changes in the liver tissue of rainbow trout (*Oncorhynchus mykiss*) exposed to sub-lethal concentrations of diazinon. Fish Physiology and Biochemistry. 2013;**39**(3):489-501. DOI: 10.1007/s10695-012-9714-1

[3] Goode ABC, Tipping PW, Gettys LA, Knowles BK, Pokorny E, Salinas LS. Integrating herbicide rates, coverage, and classical biological control insects (*Megamelus scutellaris, Neochetina eichhorniae*, and *Neochetina bruchi*) to manage Pontederia (Eichhornia) crassipes. Biological Control. 2022;**170**:104930. DOI: 10.1016/j.biocontrol.2022.104930

[4] Reichert LMM, de Oliveira DR, Papaleo JL, Valgas AAN, Oliveira GT. Biochemical and body condition markers in Rhinella icterica tadpoles exposed to atrazine, glyphosate, and quinclorac based herbicides in ecologically relevant concentrations. Environmental Toxicology and Pharmacology. 2022;**93**:103884. DOI: 10.1016/j.etap.2022.103884

[5] James TK, Ghanizadeh H, Harrington KC, Bolan NS. The leaching behaviour of herbicides in cropping soils amended with forestry biowastes. Environmental Pollution. 2022;**307**:119466. DOI: 10.1016/j.envpol.2022.119466

[6] Bai SH, Ogbourne SM. Glyphosate: Environmental contamination, toxicity and potential risks to human health via food contamination. Environmental Science and Pollution Research. 2016;**23**(19):18988-19001. DOI: 10.1007/s11356-016-7425-3

[7] Sun Y, Cao M, Wan Y, Wang H, Liu J, Pan F, et al. Spatial variation of 2,4-D and MCPA in tap water and groundwater from China and their fate in source, treated, and tap water from Wuhan, Central China. Science and Total Environment. 2020;**727**:138691. DOI: 10.1016/j.scitotenv.2020.138691

[8] Correia NM, Carbonari CA, Velini ED. Detection of herbicides in water bodies of the samambaia river sub-basin in the federal district and eastern Goiás. Journal of Environmental Science and Health, Part B. 2020;**55**:574-582. DOI: 10.1080/03601234.2020.1742000

[9] Rimayi C, Odusanya D, Weiss JM, de Boer J, Chimuka L. Seasonal variation of chloro-s-triazines in the Hartbeespoort dam catchment, South Africa. Science of the Total Environment. 2018;**613-614**:472-482. DOI: 10.1016/j.scitotenv.2017.09.119

[10] Tyohemba RL, Pillay L, Humphries MS. Bioaccumulation of current-use herbicides in fish from a global biodiversity hotspot: Lake St. Lucia, South Africa. Chemosphere. 2021;**284**:131407. DOI: 10.1016/j.chemosphere.2021.13140

[11] Ojemaye CY, Onwordi CT, Pampanin DM, Sydnes MO, Petrik L. Presence and risk assessment of herbicides in the marine environment of Camps Bay (Cape Town, South Africa). Sci. Total Environ. 2020;**738**:140346. DOI: 10.1016/j.scitotenv.2020.140346

[12] Rodrigues ET, Alpendurada AF, Ramos F, Pardal MÂ. Environmental and human health risk indicators for agricultural pesticides in estuaries. Ecotoxicology and Environmental Safety. 2018;**150**:224-231. DOI: 10.1016/j.ecoenv.2017.12.047

[13] Ahmadi K, Mirvaghefei AR, Banaee M, Vosoghei AR. Effects of long-term diazinon exposure on some immunological and haematological parameters in rainbow trout *Oncorhynchus mykiss* (Walbaum, 1792). Toxicology and Environmental Health Sciences. 2014;**6**:1-7. DOI: 10.1007/s13530-014-0181-1

[14] Banaee M, Soltanian S, Sureda A, Gholamhosseini A, Haghi BN, Akhlaghi M, et al. Evaluation of single and combined effects of cadmium and micro-plastic particles on biochemical and immunological parameters of common carp (*Cyprinus carpio*). Chemosphere. 2019;**236**:124335. DOI: 10.1016/j.chemosphere.2019.07.066

[15] Derikvandy A, Pourkhabbaz HR, Banaee M, Sureda A, Haghi N, Pourkhabbaz AR. Genotoxicity and oxidative damage in zebrafish (*Danio rerio*) after exposure to effluent from ethyl alcohol industry. Chemosphere. 2020;**251**:126609. DOI: 10.1016/j.chemosphere.2020.126609

[16] Banaee M, Mohammadipour S, Madhani S. Effects of sublethal concentrations of permethrin on bioaccumulation of cadmium in zebra cichlid (*Cichlasoma nigrofasciatum*). Toxicological and Environmental Chemistry. 2015;**97**(2):200-207. DOI: 10.1080/02772248.2015.1031668

[17] Banaee M, Shahafve S, Vaziriyan M, Taheri S, Nemadoost B, Haghi. Effects of sewage effluent on blood biochemical parameters of common carp (*Cyprinus carpio*): A case study of Behbahan, Khuzestan Province. Journal of Chemical Health Risks. 2016;**6**(3):161-173. DOI: 10.22034/jchr.2016.544144

[18] Banaee M, Tahery S, Vaziriyan M, Shahafve S, Nemadoost-Haghi B. Reproductive health indicators of immature common carp exposed to municipal wastewater of Behbahan, Iran. Journal of Advances in Environmental Health Research. 2015;**3**(3):164-171. DOI: 10.22102/jaehr.2015.40199

[19] Banaee M, Sureda A, Taheri S, Hedayatzadeh F. Sub-lethal effects of dimethoate alone and in combination with cadmium on biochemical parameters in freshwater snail, galba truncatula. Comparative Biochemistry and Physiology Part C: Toxicology and Pharmacology. 2019;**220**:62-70. DOI: 10.1016/j.cbpc.2019.03.002

[20] Banaee M, Tahery S, Nematdoost Haghi B, Shahafve S, Vaziriyan M. Blood biochemical changes in common carp (*Cyprinus carpio*) upon co-exposure to titanium dioxide nanoparticles and paraquat. Iranian Journal of Fisheries Sciences. 2019;**18**(2):242-255. DOI: 10.22092/ijfs.2019.118174

[21] Banihashemi EA, Soltanian S, Gholamhosseini A, Banaee M. Effect of microplastics on yersinia ruckeri infection in rainbow trout (*Oncorhynchus mykiss*). Environmental Science and Pollution Research. 2022;**29**(8):11939-11950. DOI: 10.1007/s11356-021-16517-3

[22] Nematdoost Haghi B, Banaee M. Effects of micro-plastic particles on paraquat toxicity to common carp (*Cyprinus carpio*): Biochemical changes. International journal of Environmental Science and Technology. 2017;**14**(3):521-530. DOI: 10.1007/s13762-016

[23] Tabche LM, Arias DG, Hidalgo ES, Galar CI. Toxic effects of Paraquat and lead on fish liver (Oreochfomis hornorum). European Journal of Pharmacology. 1990;**183**(4):1534-1535. DOI: 10.1016/0014-2999(90)94691-P

[24] Xu X, Cui Z, Wang X, Wang X, Zhang S. Toxicological responses on cytochrome P450 and metabolic transferases in liver of goldfish (Carassius auratus) exposed to lead and paraquat. Ecotoxicology and Environmental Safety. 2018;**151**:161-169. DOI: 10.1016/j.ecoenv.2017.12.062

[25] de Lima Faria JM, Guimarães LN, da Silva VC, de Oliveira Lima EC, de Sabóia-Morais SMT. Recovery trend to co-exposure of iron oxide nanoparticles (γ-Fe2O3) and glyphosate in liver tissue of the fish *Poecilia reticulata*. Chemosphere. 2021;**282**:130993. DOI: 10.1016/j.chemosphere.2021.130993

[26] Banaee M, Akhlaghi M, Soltanian S, Gholamhosseini A, Heidarieh H, Fereidouni MS. Acute exposure to chlorpyrifos and glyphosate induces changes in hemolymph biochemical parameters in the crayfish, *Atacus leptodactylus* (Eschscholtz, 1823). Comparative Biochemistry and Physiology Part C: Toxicology and Pharmacology. 2019;**222**:145-155. DOI: 10.1016/j.cbpc.2019.05.003

[27] Banaee M, Akhlaghi M, Soltanian S, Sureda A, Gholamhosseini A, Rakhshaninejad M. Combined effects of exposure to sub-lethal concentration of the insecticide chlorpyrifos and the herbicide glyphosate on the biochemical changes in the freshwater crayfish *Pontastacus leptodactylus*. Ecotoxicology. 2020;**29**(9):1500-1515. DOI: 10.1007/s10646-020-02233-0

[28] Bonifacio AF, Zambrano MJ, Hued AC. Integrated ecotoxicological assessment of the complex interactions between chlorpyrifos and glyphosate on a non-target species *Cnesterodon decemmaculatus* (Jenyns, 1842). Chemosphere. 2020;**261**:127782. DOI: 10.1016/j.chemosphere.2020.127782

[29] Banaee M. Physiological dysfunction in fish after insecticides exposure. In: Trdan S, editor. Insecticides—Development of Safer and More Effective Technologies. London: InTech; 2013. pp. 103-143. DOI: 10.5772/54742

[30] Nassar AMK, Abdel-Halim KY, Abbassy MA. Mitochondrial biochemical and histopathological defects induced by the herbicide pendimethalin in tilapia fish (*Oreochromis niloticus*). Comparative Biochemistry and Physiology Part C: Toxicology & Pharmacology. 2021;**242**:108949. DOI: 10.1016/j.cbpc.2020.108949

[31] Zhang X, Ivantsova E, Perez-Rodriguez V, Cao F, Souders CL, Martyniuk CJ. Investigating mitochondria-immune responses in zebrafish, Danio rerio (Hamilton, 1822): A case study with the herbicide dinoseb. Comparative Biochemistry and Physiology Part C: Toxicology & Pharmacology. 2022;**257**:109357. DOI: 10.1016/j.cbpc.2022.109357

[32] Sharifinasab Z, Banaee M, Mohiseni M, Noori A. vitamin C and chitosan alleviate toxic effects of Paraquat on some biochemical parameters in hepatocytes of common carp. Iranian Journal of Toxicology. 2016;**10**(1):31-40. DOI: doi

[33] Pereira L, Fernandes MN, Martinez CBR. Hematological and biochemical alterations in the fish Prochilodus lineatus caused by the herbicide clomazone. Environmental Toxicology and Pharmacology.

2013;**36**(1):1-8. DOI: 10.1016/j.etap.2013.02.019

[34] Piancini LDS, Guiloski IC, de Assis HCS, Cestari MM. Mesotrione herbicide promotes biochemical changes and DNA damage in two fish species. Toxicology Reports. 2015;**2**:115-1163. DOI: 10.1016/j.toxrep.2015.08.007

[35] Destro ALF, Silva SB, Gregório KP, de Oliveira JM, Lozi AA, Zuanon JAS, et al. Effects of subchronic exposure to environmentally relevant concentrations of the herbicide atrazine in the Neotropical fish *Astyanax altiparanae*. Ecotoxicology and Environmental Safety. 2021;**208**:111601. DOI: 10.1016/j.ecoenv.2020.111601

[36] Moraes BS, Loro VL, Pretto A, da Fonseca MB, Menezes C, Marchesan E, et al. Toxicological and metabolic parameters of the teleost fish (*Leporinus obtusidens*) in response to commercial herbicides containing clomazone and propanil. Pesticide Biochemistry and Physiology. 2009;**95**(2):57-62. DOI: 10.1016/j.pestbp.2009.06.006

[37] Sinhorin VD, Sinhorin AP, Teixeira JM, Miléski KM, Hansen PC, Moreira PS, et al. Effects of the acute exposition to glyphosate-based herbicide on oxidative stress parameters and antioxidant responses in a hybrid Amazon fish Surubim (Pseudoplatystoma sp). Ecotoxicology and Environmental Safety. 2014;**106**:181-187. DOI: 10.1016/j.ecoenv.2014.04.040

[38] Thanomsit C, Kiatprasert P, Prasatkaew W, Khongchareonporn N, Nanthanawat P. Acetylcholinesterase (AChE) monoclonal antibody generation and validation for use as a biomarker of glyphosate-based herbicide exposure in commercial freshwater fish. Comparative Biochemistry and Physiology Part C: Toxicology & Pharmacology. 2021;**241**:108956. DOI: 10.1016/j.cbpc.2020.108956

[39] Solomon KR, Dalhoff K, Volz D, Kraak GVD. Effects of herbicides on fish. In: Farrell AP, Brauner CJ, Tierney KB, editors. Fish Physiology. Elsevier, Academic Press. 2013. pp. 369-409. DOI: 10.1016/B978-0-12-398254-4.00007-8

[40] Geetha N. Mitigatory role of butyrylcholinesterase in freshwater fish Labeo rohita exposed to glyphosate based herbicide roundup®. Materials Today: Proceedings. 2021;**47**(9):2030-2035. DOI: 10.1016/j.matpr.2021.04.281

[41] Ruiz de Arcaute C, Soloneski S, Larramendy ML. Evaluation of the genotoxicity of a herbicide formulation containing 3,6-dichloro-2-metoxybenzoic acid (dicamba) in circulating blood cells of the tropical fish Cnesterodon decemmaculatus. Mutation Research, Genetic Toxicology and Environmental Mutagenesis. 2014;**773**:1-8. DOI: 10.1016/j.mrgentox.2014.08.001

[42] Guilherme S, Santos MA, Gaivão I, Pacheco M. Genotoxicity evaluation of the herbicide Garlon(®) and its active ingredient (triclopyr) in fish (*Anguilla anguilla* L.) using the comet assay. Environmental Toxicology. 2015;**30**(9):1073-1081. DOI: 10.1002/tox.21980

[43] Cavalcante DG, Martinez CB, Sofia SH. Genotoxic effects of roundup on the fish *Prochilodus lineatus*. Mutation Research. 2008;**655**(1-2):41-46. DOI: 10.1016/j.mrgentox.2008.06.010

[44] Cavas T. In vivo genotoxicity evaluation of atrazine and atrazine-based herbicide on fish *Carassius auratus* using the micronucleus test and the comet assay. Food and Chemical Toxicology. 2011;**49**(6):1431-1435. DOI: 10.1016/j.fct.2011.03.038

[45] Vera-Candioti J, Soloneski S, Larramendy ML. Single-cell gel electrophoresis assay in the ten spotted live-bearer fish, Cnesterodon decemmaculatus (Jenyns, 1842), as bioassay for agrochemical-induced genotoxicity. Ecotoxicology and Environmental Safety. 2013;**98**:368-373. DOI: 10.1016/j.ecoenv.2013.08.011

[46] Bernardi F, Lirola JR, Cestari MM, Bombardelli RA. Effects on reproductive, biochemical and genotoxic parameters of herbicides 2,4-D and glyphosate in silver catfish (*Rhamdia quelen*). Environmental Toxicology and Pharmacology. 2022;**89**:103787. DOI: 10.1016/j.etap.2021.103787

[47] Velki M, Meyer-Alert H, Seiler TB, Hollert H. Enzymatic activity and gene expression changes in zebrafish embryos and larvae exposed to pesticides diazinon and diuron. Aquatic Toxicology. 2017;**193**:187-200. DOI: 10.1016/j.aquatox.2017.10.019

[48] Smith CM, Vera MKM, Bhandari RK. Developmental and epigenetic effects of roundup and glyphosate exposure on Japanese medaka (Oryzias latipes). Aquatic Toxicology. 2019;**210**:215-226. DOI: 10.1016/j.aquatox.2019.03.005

[49] Groff JM. Neoplasia in fishes. The Veterinary Clinics of North America. Exotic Animal Practice. 2004;**7**(3):705, vii-756. DOI: 10.1016/j.cvex.2004.04.012

[50] Ma J, Zhu J, Wang W, Ruan P, Rajeshkumar S, Li X. Biochemical and molecular impacts of glyphosate-based herbicide on the gills of common carp. Environmental Pollution. 2019;**252B**:1288-1300. DOI: 10.1016/j.envpol.2019.06.040

[51] Maddalon A, Galbiati V, Colosio C, Mandić-Rajčević S, Corsini E. Glyphosate-based herbicides: Evidence of immune-endocrine alteration. Toxicology. 2021;**459**:152851. DOI: 10.1016/j.tox.2021.152851

[52] Acar Ü, İnanan BE, Navruz FZ, Yılmaz S. Alterations in blood parameters, DNA damage, oxidative stress and antioxidant enzymes and immune-related genes expression in Nile tilapia (*Oreochromis niloticus*) exposed to glyphosate-based herbicide. Comparative Biochemistry and Physiology Part C: Toxicology & Pharmacology. 2021;**249**:109147. DOI: 10.1016/j.cbpc.2021.109147

[53] Zebral YD, Lansini LR, Costa PG, Roza M, Bianchini A, Robaldo RBA. Glyphosate-based herbicide reduces fertility, embryonic upper thermal tolerance and alters embryonic diapause of the threatened annual fish Austrolebias nigrofasciatus. Chemosphere. 2018;**196**:260-269. DOI: 10.1016/j.chemosphere.2017.12.196

[54] Yusof S, Ismail A, Alias MS. Effect of glyphosate-based herbicide on early life stages of Java medaka (Oryzias javanicus): A potential tropical test fish. Marine Pollution Bulletin. 2014;**85**(2):494-498. DOI: 10.1016/j.marpolbul.2014.03.022

[55] Dehnert GK, Karasov WH, Wolman MA. 2,4-Dichlorophenoxyacetic acid containing herbicide impairs essential visually guided behaviors of larval fish. Aquatic Toxicology. 2019;**209**:1-12. DOI: 10.1016/j.aquatox.2019.01.015

[56] Dehnert GK, Freitas MB, Sharma PP, Barry TP, Karasov WH. Impacts of subchronic exposure to a commercial 2,4-D herbicide on developmental stages of multiple freshwater fish species. Chemosphere. 2021;**263**:127638. DOI: 10.1016/j.chemosphere.2020.127638

[57] Merola C, Fabrello J, Matozzo V, Faggio C, Iannetta A, Tinelli A, et al. Dinitroaniline herbicide pendimethalin affects development and induces biochemical and histological alterations in zebrafish early-life stages. Science of the Total Environment. 2022;**828**:154414. DOI: 10.1016/j.scitotenv.2022.154414

[58] Zhang C, Wang Z, Liu S, Tan H, Zeng D, Li X. Analytical method for sequential determination of persistent herbicides and their metabolites in fish tissues by UPLC–MS/MS. Chemosphere. 2022;**288**(2):132591. DOI: 10.1016/j.chemosphere.2021.132591

[59] Peng J, Gan J, Ju X, Liu T, Chen J, He L. Analysis of triazine herbicides in fish and seafood using a modified QuEChERS method followed by UHPLC-MS/MS. Journal of Chromatography. B, Analytical Technologies in the Biomedical and Life Sciences. 2021;**1171**:122622. DOI: 10.1016/j.jchromb.2021.122622

[60] Sun X, Gao J, Xing J, Xing L, Guo M, Peng J, et al. Simultaneous determination of triazine herbicides and their metabolites in shellfish by HPLC-MS/MS combined with Q/E-Orbitrap HRMS. Analytical and Bioanalytical Chemistry. 2021;**413**:6239-6252. DOI: 10.1007/s00216-021-03579-y